T0339124

Why Free Will Is Real

Why Free Will Is Real

Christian List

Harvard University Press
Cambridge, Massachusetts & London, England | 2019

Printed in the United States of America

Second printing

Library of Congress Cataloging-in-Publication Data

Names: List, Christian, author.
Title: Why free will is real / Christian List.
Description: Cambridge, Massachusetts : Harvard University Press, 2019. |
Includes bibliographical references and index.
Identifiers: LCCN 2018046432 | ISBN 9780674979581 (hardcover : alk. paper)
Subjects: LCSH: Free will and determinism. | Physical laws. | Physics—Religious aspects.
| Religion and science.
Classification: LCC BJ1461 .L55 2019 | DDC 123/.5—dc23 LC record available at
https://lccn.loc.gov/2018046432

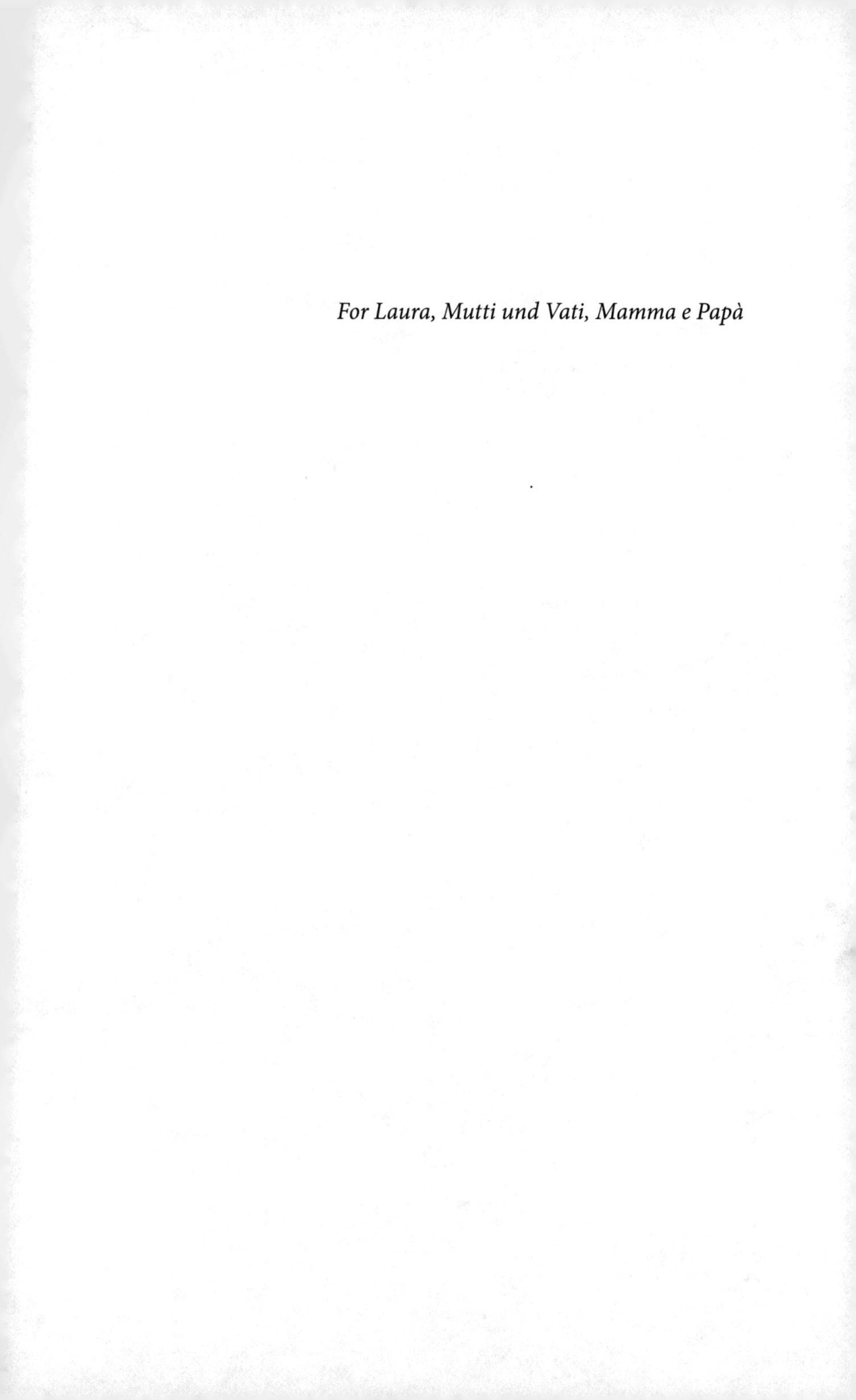

For Laura, Mutti und Vati, Mamma e Papà

Contents

Introduction 1

1. Free Will 15

2. Three Challenges 31

3. In Defence of Intentional Agency 49

4. In Defence of Alternative Possibilities 79

5. In Defence of Causal Control 113

Conclusion 149

Notes *161*

References *187*

Acknowledgments *203*

Index *207*

Introduction

Free will *is* an illusion. Our wills are simply not of our own making. Thoughts and intentions emerge from background causes of which we are unaware and over which we exert no conscious control. We do not have the freedom we think we have.

—Sam Harris, neuroscientist, in *Free Will*

Our thoughts and actions are the outputs of a computer made of meat—our brain—a computer that must obey the laws of physics. Our choices, therefore, must also obey those laws. This puts paid to the traditional idea of . . . free will: that our lives comprise a series of decisions in which *we could have chosen otherwise*. . . . So why does the term "free will" still hang around when science has destroyed its conventional meaning? Some . . . are impressed by their feeling that they *can* choose, and must comport this with science. Others have said explicitly that characterizing "free will" as an illusion will hurt society. If people believe they're puppets, well, then maybe they'll be crippled by nihilism, lacking the will to leave their beds. This attitude reminds me of the (probably apocryphal) statement of the Bishop of Worcester's wife when she heard about Darwin's theory: "My dear, descended from the apes! Let us hope it is not true, but if it is, let us pray it will not become generally known."

—Jerry Coyne, biologist, in
What Scientific Idea Is Ready for Retirement?

The idea of free will is as old as it is entrenched in our thinking. As far back as the Old Testament we find the idea that human beings are capable of making choices for which they can be held responsible. Think of Adam and Eve's choice to eat the forbidden fruit. Or think of Abraham's decision whether to sacrifice his son Isaac. In modern societies, free will is central to our practices of praising and blaming one another and of attributing responsibility. Could we hold someone responsible for something they did not freely choose? Would blaming them be warranted? Try to imagine what the criminal justice system would look like if society accepted the view that people's choices are not up to them. How could a court find someone criminally liable for an action they did not freely choose? Without free will, the familiar notion of responsibility would be in trouble.

At a more personal level, free will is at the heart of our self-conception as agents capable of deliberating about our actions. Unless our choices are up to us, there is little point in asking "What should I do?" When we deliberate about how to act—what career path to take, what projects to pursue, whom to marry, whether to help someone in need—we must recognize the different options as genuine possibilities among which we can choose. Otherwise, deliberation would be pointless. Our sense of agency is inextricably bound up with the idea that we are capable of making real choices, at least in principle.

Yet the idea of free will is under attack. Philosophers have long debated whether there could be free will if the world was deterministic. "Determinism" is the notion that the past fully determines the future: once the initial state of the world is given—say, at the time of the Big Bang—all subsequent events unfold in a predetermined sequence, governed by the laws of nature, as in a mechanical clockwork. The idea of a deterministic universe is familiar from the worldview of the Enlightenment. It is associated with thinkers such as Isaac Newton, Pierre Laplace, and, in the twentieth century, Albert Einstein. In a deterministic universe, it seems, there would be no room for free choices. Everything we did would be predetermined long before our actions. The fact that you are reading these lines now would be an inevitable consequence of the world's initial conditions at the time of the Big Bang, given the laws of nature. Similarly, the way you will vote in the next election would be no less predetermined than the date of the next total solar eclipse in the United Kingdom. (In case you are interested, it is due to take place on 23 September 2090.)

Determinism is not the only threat to free will. Even if the universe is not deterministic and the future is truly open, it does not automatically follow that our choices are up to us. They could still be flukes of randomness, or they could be caused by factors beyond our control, such as subconscious states of our brains. In recent years, an increasing number of scientists have claimed that free will does not fit into our best scientific understanding of the world. People's choices are produced by neuronal activity in their brains, in a purely physical manner, and it is unclear what role any intentional decisions are supposed to play. The opening quotations illustrate this view. If the laws of physics govern everything that happens in the world, and we are part of that world, then it is hard to see how our choices could be truly ours: "We do not have the freedom we think we have," as Sam Harris puts it.

Even some scholars in the humanities acknowledge the threat. The historian Yuval Noah Harari, for instance, writes,

> Humanism is now facing an existential challenge and the idea of "free will" is under threat. Scientific insights into the way our brains and bodies work suggest that our [thoughts and] feelings are not some uniquely human spiritual quality. Rather, they are biochemical mechanisms that all mammals and birds use in order to make decisions by quickly calculating probabilities of survival and reproduction. . . . Even though humanists were wrong to think that our feelings reflected some mysterious "free will," up until now humanism still made very good practical sense. For although there was nothing magical about our feelings, they were nevertheless the best method in the universe for making decisions. . . . Even if the Catholic Church or the Soviet KGB spied on me every minute of every day, they lacked the biological knowledge and the computing power necessary to calculate the

> biochemical processes shaping my desires and choices. . . . However, as the Church and the KGB give way to Google and Facebook [which can predict human behaviour using science and big data], humanism loses its practical advantages.[1]

In short, the picture of human beings as agents capable of making free choices is on its way out. Free will, it seems, is an ingredient of an outdated, prescientific worldview.

The aim of this book is to set out a strategy for answering these scientific challenges for free will. I want to defend a picture of free will that has largely escaped people's notice, despite all the attention the subject has received. This picture construes free will as a "higher-level" phenomenon: a phenomenon that is found not at the level of fundamental physics but at the level of psychology, and specifically at the level of intentional agents: goal-directed beings like us. Free will, I will argue, is in the company of other phenomena that emerge from the physical world but that are not best understood in fundamental physical terms themselves. Familiar examples of emergent phenomena are living organisms and ecosystems in biology, the mind in psychology, and institutions, cultures, and the market in the social realm. All these ultimately emerge from physical processes, but we need to go beyond physics to understand them. Looking at them solely through the lens of the physical laws governing particles and molecules, for instance, would give us little insight.

Free-will sceptics typically start from the premise that free will requires property *P*, where *P* might be one or several of the following:

- intentional, goal-directed agency;
- alternative possibilities among which we can choose;
- causation of our actions by our mental states, specifically by our intentions.

But then they claim that physics, or some other fundamental science, shows that there is no such thing as *P*: *P* is not to be found among the fundamental physical features of the world. Perhaps it is just a convenient fiction of our prescientific way of thinking—a "folk idea." So, if *P* is required for free will, then, science seems to tell us, there is no free will.[2] Different arguments against free will target different substitution instances for *P*. Some suggest that intentional agency is an illusion. Others assert that we could not have alternative possibilities in a deterministic universe. Still others insist that our intentions never cause our actions: when I act, it is my brain that makes me do it.[3] What all these arguments have in common is that they take the relevant property *P* to be out of place in a scientific worldview.

My response is that it may well be true that there is no such thing as *P at the fundamental physical level*. But this observation does not show that there is no such thing as *P* at all. Free will and its prerequisites—such as intentional agency, alternative possibilities, and causal control over our actions—are *higher-level phenomena*. Yet they are no less real for that. If we are looking for free will at the physical level, we are simply looking in the wrong place. In the chapters that follow, I will develop this response in detail. For now I would like to summarize the key idea.

Let me begin with the notion of *levels*.[4] Facts about the world—in general, not just in relation to human agency and free will—are stratified into levels. In the sciences, "levels" correspond to different ways we describe the world. To make sense of the basic laws of nature, for example, we employ fundamental physical descriptions, drawing on our best theories of physics. We use concepts such as particles, fields, and forces, and specify a variety of equations that describe how physical systems evolve over time. By contrast, to make sense of chemical or biological phenomena, we need to go beyond fundamental physics. Molecules, cells, and organisms all display patterns and regularities that can only be captured at another level of description, using a conceptual repertoire distinct from that of fundamental physics. As philosophers of science have pointed out, even a property as simple as acidity cannot be satisfactorily described in fundamental physical terms alone.[5] In the case of acidity, there is no easily describable configuration of fundamental physical properties that exactly matches this chemical property. In particular, there is no "translation scheme" that fully translates talk of acidity into talk of particles, fields, and forces alone. We need the concepts and categories of chemistry to talk about acidity. And we are here still dealing with a fairly basic example, compared to other, more complex chemical or biological phenomena.

In case you are in doubt about the need to go beyond fundamental physics to make sense of the world, try to explain cell division, genetic inheritance, or evolution by referring to nothing but molecules, atoms, and other elementary particles. Each living cell consists of billions or trillions of atoms, and an organism consists of billions or trillions of cells.[6] Even the best supercomputer would struggle with the astronomical task of computing the processes inside a single organism at the atomic or molecular level. And even if, against all odds, these difficulties could be overcome, perhaps with the help of massively distributed computing on the internet, then a purely microphysical description of cell division, genetic inheritance, or evolution would still fail to pick up many macroscopic regularities we are interested in. Indeed, such a description would not help us to understand the relevant phenomena at all. It would make us lose sight of the forest for the trees. Now, once we turn to the domain of humans and their intentional actions, fundamental physical descriptions are wholly inadequate. In the language of fundamental physics, we cannot even talk about tables, trees, and other ordinary objects—only about particles, fields, forces, and so on. If we wish to make sense of people and what they do, we require psychological descriptions—descriptions that refer to thoughts and beliefs, preferences and desires, goals and intentions.

I will use the term "lower-level descriptions" to refer to descriptions at the fundamental physical level or at some other microlevel—say, that of molecular chemistry. And I will use the term "higher-level descriptions" to refer to descriptions at a more macroscopic level, as in biology, psychology, or sociology. I will call any phenomenon that can be adequately captured only by higher-level descriptions a "higher-level phenomenon." Thus, higher-level descriptions are markers of higher-level phenomena. Higher-level phenomena are ubiquitous; they permeate much of our human environment. Anything ranging from DNA to the weather falls into this category, as does the phenomenon of intentional agency itself, together with the associated phenomena of thought, belief, desire, intention, and choice. Of course, "lower level" and "higher level" are relative notions. Neuroscience, for example, operates at a lower level than cognitive psychology, but at a higher level than fundamental physics.[7]

Against this background it will become clear that free will is very much a *higher-level* phenomenon. For this reason, we should not be surprised if a person's capacity to make free choices is nowhere to be found at the physical level, or even at the neurophysiological level. What our inability to find freedom at that level shows is merely that free will is not a physical or neurophysiological phenomenon; it does not show that it is unreal.

Of course, higher-level phenomena are not free-floating. Any higher-level phenomenon depends on what goes on at the physical level: it "supervenes" on it, as philosophers say.[8] When a chemical reaction happens, a cell divides, or a person performs an action, this is ultimately the result of some underlying physical process. Without that physical process, there would be no chemical reaction, no cell division, and no human action. This is what the scientific worldview tells us. At bottom, all higher-level phenomena stem from the interaction of elementary particles, fields, and forces.

However—and this is important—although higher-level phenomena supervene on fundamental physical phenomena, they are no less real for that. The economy, for example, is ultimately produced by underlying physical processes; without these processes, there would be no life on this planet, and without life, there would be no economy. Yet, we would never think that phenomena such as the interest rate, inflation, and unemployment are unreal. It would be a mistake, for instance, to overlook the causal connection between monetary policy and inflation, and it would be absurd to suggest that, just because unemployment is a higher-level phenomenon, it must be unreal, or somehow less real than kinetic energy or electromagnetism. The bottom line is that higher-level phenomena stand in regularities of their own and serve as causes and effects of other phenomena. If we did not recognize those regularities, we would be overlooking some significant aspects of the world.

In line with this, I will assume that a phenomenon qualifies as *real* if recognizing its existence is explanatorily indispensable: we would fail to give an adequate explanation of the relevant domain without recognizing the phenomenon in question. This criterion of reality is familiar from the sciences. It is rooted in what is sometimes called a "naturalistic ontological attitude": the view that the scientific method is a good guide to ontological questions—that is, questions about what does or does not exist.[9] Moreover, it would be a mistake to apply the naturalistic criterion of reality only to the domain of fundamental physics. Rather, it is equally applicable to the domains of the special sciences, from biology and geology to the human and social sciences.

The interest rate, inflation, and unemployment all meet this criterion of reality, as do ordinary large-scale objects and their various properties: we have very good scientific reasons to treat them as real. Similarly, I will show that the phenomenon of intentional, goal-directed agency passes the test. Although intentional agency is the result of physical and biological processes in a person's brain and body, we could not seriously deny its reality. How people behave is typically best understood in agential terms. People's beliefs, desires, and intentions cause them to act in certain ways, and we would fail to understand this adequately if we viewed each person as nothing but a heap of interacting particles.

Why, for instance, do people show up for work in the morning, except when they are ill, on holiday, or in an emergency? Why do people become politically active? Why do they study or engage in cultural activities? Why do some people break the law, but less so if they are likely to get caught? The answers to these questions lie in people's beliefs, desires, and intentions, as well as in the practices and incentives structuring their lives. People respond to their circumstances in more or less intelligible ways, and their choices, in turn, are affected by certain norms and conventions of their social environments. All of this can be described in a reasonably adequate way using the concepts and categories of psychology and of the social sciences, where people are viewed as goal-directed agents whose actions are guided by their intentional mental states. By contrast, if we tried to make sense of people as heaps of interacting physical particles, we would be employing the wrong level of description. Instead of understanding a person's behaviour as an intelligible and meaningful response to his or her perspective on the world, we would have to trace the astronomically complicated processes in the physical brain and body. We would miss the psychological and social regularities we are interested in.

Once we absorb the lessons of these points, we can see that intentional agency is a real phenomenon. Acknowledging its existence is not just scientifically respectable, but it is indispensable for many explanatory purposes. Intentional agency is therefore no less real than other macroscopic phenomena recognized by the sciences, from the solidity of surfaces to the patterns of the climate. I will argue that the same is true of the other prerequisites for free will. A person's ability to choose between different courses of action—alternative possibilities—is a real phenomenon, as is the person's causal control over his or her actions.

Putting all this together paves the way for a defence of free will as a higher-level phenomenon—and crucially, a defence that is compatible with a scientific worldview. I will call the resulting picture of free will "compatibilist libertarianism":

- "compatibilist" because of its compatibility with science, including physical determinism; and
- "libertarianism" because of its commitment to the idea that free will involves a genuine ability to choose between different actions.[10]

I should, however, make a further remark about this terminology. While the term "compatibilist" should not give rise to any misunderstandings, the term "libertarianism" can unfortunately have two different meanings. In debates about free will, it refers to those philosophical views that take free will to be real and to involve alternative possibilities. This is the intended meaning here. In political philosophy, by contrast, the same term is commonly employed to denote certain *political* views according to which liberty is the central political value—views associated with thinkers such as John Locke and Robert Nozick. Libertarianism about free will does not entail libertarianism in the political sense and should not be confused with it. Throughout this book, I will use the terms "libertarianism" and "libertarian" solely in their nonpolitical sense. Readers who do not like the label "compatibilist libertarianism" might use the label "free-will emergentism" instead. It also highlights what is distinctive about the picture of free will that I defend.[11]

My argument for free will is structured as follows. In Chapter 1, I will introduce the idea of free will in more detail and discuss what I consider the three jointly necessary and sufficient requirements for free will: intentional agency, alternative possibilities, and causal control over one's actions. These requirements, I will suggest, express what it takes to have free will in a fairly conventional sense. In Chapter 2, I will describe three scientifically motivated challenges for free will—one for each requirement—which purport to show that the requirements cannot be met if the world is as depicted by science. I will call these "the challenge from radical materialism," "the challenge from determinism," and "the challenge from epiphenomenalism." In Chapters 3, 4, and 5, I will then present my defence of free will, devoting one chapter to each challenge. In these chapters, I will not only explain where each of the challenges goes wrong, but also make a positive case for free will. I will conclude the book with some general remarks about the resulting picture of free will.

The task I have set for myself is ambitious, and I should emphasize the book's limitations. First, I cannot deliver a full-fledged theory of free will. I can only provide an account of the central ideas, all the more so because I will keep the book relatively brief. Developing a comprehensive theory of free will goes beyond the scope of any single piece of work and can only be achieved by a community of scholars, across multiple disciplines.

Second, the picture of free will that I present depends on certain empirical premises. Some of my claims will be correct only if those premises are true. And demonstrating their truth is beyond the scope of this book. Science will ultimately have to adjudicate them. On the positive side, this means that my account *could* be wrong. If someone asks, "What would it take for this account of free will to be wrong?," I do have an answer. Like other scientific theories, my account meets a kind of falsifiability requirement.

Third, I will focus on the *architecture* of what I take to be the correct account of free will, not the details. My central thesis, as already explained, is that the key to reconciling free will with a scientific worldview lies in the distinction between phenomena at different levels. I will explain how drawing that distinction allows us to make progress in our understanding of free will, but there may be other ways in which we could develop a theory of free will around this architecture. My key claim—that free will is a higher-level phenomenon—is compatible with a number of different ways of spelling out the details.

Finally, I will approach the subject of free will from a third-person perspective—the perspective of an external observer investigating what it is for an agent to have free will—rather than from a first-person perspective—the internal perspective of the bearer of free will him- or herself. Much of science shares this third-person perspective, though not all of philosophy does. Phenomenology, for example, is an approach to philosophy that investigates how we experience the world from a first-person perspective. Perhaps a complete theory of free will must analyze this phenomenon not just from the outside but also from the inside. And there is clearly an important question of how free will is related to first-person consciousness. I will have to set this question aside, though I hope that my analysis will be consistent with future, first-person investigations.

In developing my arguments, I stand on the shoulders of others. The literature on free will is large and sophisticated, and hardly any ideas in this area are ever completely new. Indeed, for a long time, I was reluctant to write this book, as I was daunted by the wealth of existing work on the topic.[12] Since 2011, I have, however, presented earlier versions of my ideas in a number of articles and lectures, including joint articles with Peter Menzies and Wlodek Rabinowicz.[13] The feedback I have received has encouraged me to pursue these ideas further.

Important precursors to my account of free will include the works of Anthony Kenny, who in the 1970s defended free will by invoking a distinction between the physiological and psychological levels and argued that determinism at the physiological level does not rule out free will at the psychological one;[14] and the works of Daniel Dennett, who argued—especially in a 2003 book—that, like the phenomenon of life itself, "freedom evolves," even if the underlying physics is deterministic.[15] And as early as the 1960s A. I. Melden suggested that physical descriptions do not adequately capture the phenomenon of free action, though he did not precisely explain how free action fits into the physical world.[16]

My claim that the mistake in some arguments against free will lies in a conflation of levels—the physical and agential levels—further echoes Mark Siderits's suggestion in a 2008 article that "the illusion of incompatibilism only arises when we illegitimately mix two distinct vocabularies, one concerned with persons, the other concerned with the parts to which persons are reducible." Siderits describes his approach as "paleo-compatibilist."[17]

In recent years, philosophers such as Mark Balaguer, Carl Hoefer, Jenann Ismael, Robert Kane, Alfred Mele, Eddy Nahmias, Adina Roskies, and Helen Steward have done much to advance our understanding of free will from a scientifically informed perspective, and the approach I am taking shares some features with each of these scholars' works.[18] I am also not alone in attempting to combine libertarian and compatibilist ideas about free will. Others who have done so, though in different ways, include Kadri Vihvelin, who defends an account of free will she calls "libertarian compatibilism"; Helen Beebee and Alfred Mele, who argue that a particular form of compatibilism inspired by David Hume is similar in some respects to libertarianism; and Bernard Berofsky, another defender of a Humean version of compatibilism that takes libertarian intuitions seriously.[19] And finally, my approach is in line with the views of some nonreductionistically minded physicists.[20]

These other works notwithstanding, each time I gave a lecture on free will or received comments on my articles, I found that my ideas were considered more provocative and novel than I had expected. And so I have decided that it is worth presenting my ideas in the form of a book. I will write it in a more informal, less technical, and less hedged style than a scholarly article, in the hope that it will be thought provoking and reasonably accessible to nonspecialists.

1
Free Will

What Is Free Will?

Free will, on a first gloss, is an agent's capacity to choose and control his or her own actions. Sometimes this is also called "freedom of action" or "freedom of choice," but I will use "free will" as the conventional term. According to the Oxford Dictionaries website, it is "the power of acting without the constraint of necessity or fate; the ability to act at one's own discretion."[1] Common sense suggests that, as human beings, we all have this capacity. When you go to a café, it is your free choice to have one kind of drink (say, coffee) rather than another (say, tea or orange juice). Similarly, if you have time tonight, it will be your free choice whether to switch on the radio or not. More significantly, we think that our own free will is involved, at least to some extent, when we choose a partner, a career, or a travel destination. In each of these examples, there is a sense in which our choice is at least partly up to us, and we could have chosen otherwise. Of course, our choices are often constrained by our means and resources, our social context, and our history and commitments. Choosing otherwise may be costly, sometimes too costly to be feasible in practice. Think of the resident of an authoritarian country who considers criticizing the regime but then refrains from doing so for fear of retribution. Notwithstanding this, there is some leeway for free will in many situations. Even the most rigid circumstances leave our choices open at least in trivial matters: whether to sleep on our left side or our right, whether to walk a few additional steps, or whether to drink an additional glass of water. Of course, we would like our freedom to go well beyond such trivial choices.

To get a feel for the idea of free will, try the following. In a few seconds' time, perform a specific movement—say, the movement of a finger, a leg, or an eyelid. Decide which movement you wish to perform. And now do it. Were you able to exercise your free will? Common sense suggests that you were. All of the following seem to be true:

- You intended to perform the movement.
- You could have chosen a different movement, or no movement at all.
- What you did was under your control.

However trivial this little experiment may seem, it illustrates free will in a nutshell. Free will can be understood as a three-part capacity, as has been noted by other philosophers.[2] It consists of

- the capacity to act intentionally;
- the capacity to choose between alternative possibilities; and
- the capacity to control one's actions.

I will make this more precise soon. But we can already observe one thing. When we go about our daily lives, we can't suppress the powerful intuition that we all have this capacity. In the absence of severe medical or psychological impairments, the three-part capacity I have described is central to our status as agents. We have this capacity even if its exercise is sometimes constrained by the environment.

Psychological studies document the robustness of our intuitions about free will. A recent study has found that children as young as four to six years of age already have free-will intuitions resembling those of adults. By age four they recognize their own and others' ability to choose between alternative possibilities in the absence of certain physical constraints—that is, when they choose one particular action, they could also have chosen another. By age six, children even recognize a person's freedom to act against his or her own stated desires.[3]

And although free-will intuitions in adults are not always consistent,[4] there is some tentative evidence suggesting that part of our common-sense understanding of free will is shared across different cultures. In a study with participants from Colombia, Hong Kong, India, and the United States, the majority expressed intuitions consistent with what philosophers call a "libertarian" understanding of free will.[5] Libertarianism is the view that free will requires alternative possibilities, and that the world actually offers such alternative possibilities: we do indeed have genuine choices. Another study adds further nuance to this by showing that preschool-age children in both the United States and Nepal recognize the existence of freedom of choice, but that Nepalese children take social obligations to place greater constraints on action than American children do. Nadia Chernyak and colleagues conclude that "basic notions of free choice are universal," but "recognitions of social obligations as constraints on action may be culturally learned."[6]

The bottom line is that the picture of the human agent as a free chooser, capable of exercising control over his or her actions, is as common as it is deeply entrenched in our thinking.[7]

Why Does Free Will Matter?

Free will is a key presupposition of our activity of deliberation and decision making and central to our self-conception as agents. It would be difficult to ask oneself in earnest, "What should I do?," without presupposing that one is free to choose. When you weigh different courses of action against one another, you have to entertain each of them as a real possibility that you could genuinely pursue. Otherwise your deliberation would seem pointless, as your actions would not really be up to you. Along similar lines, Immanuel Kant thought that we must *presuppose* free will if we are to understand ourselves as rational and moral beings:

> Freedom must be presupposed as a property of the will of all rational beings.
>
> Reason must regard itself as the author of its principles independently of alien influences; consequently, as practical reason or as the will of a rational being it must be regarded of itself as free.[8]

Free will is also at the heart of our practices of attributing responsibility to one another, both in personal morality and in the law. We do not normally praise or blame people for things they did not do out of their own free will. Nor do we consider people capable of giving valid consent—such as when agreeing to do something—if they lack free will. In the law, a criminal offence, especially a serious one such as murder or assault, but also theft, typically "requires both a voluntary act and the intent to do wrong."[9] This, in turn, is conventionally understood to presuppose free will. Even when we blame people for their negligence and hold them liable for something without attributing intent—for example, when they have carelessly forgotten something or caused an accident through lack of attention or intoxication—the implicit assumption is that they could have taken proper care instead of being negligent. So, there is a sense in which they could have acted otherwise, at least in the run-up to what they did, for instance by adopting a more careful attitude. This presupposes that free will was present *somewhere* along the relevant chain of events.[10]

By contrast, the law has long recognized that if someone causes a harm due to circumstances beyond his or her control, such as some significant psychological or physiological condition, then he or she is not criminally liable. Although different jurisdictions differ in the kinds of excuses of this sort they accept in criminal cases, excuses by reference to "insanity," "automatism," or "irresistible impulse" are recognized in a number of jurisdictions. Of course, establishing whether someone genuinely suffers from such a responsibility-diminishing condition is difficult, and our evidence may often be inconclusive. But the conceptual point remains: the capacity of free will is widely considered necessary for fitness to be held responsible, both morally and legally.[11]

In fact, as the legal historian Thomas Andrew Green has observed, "criminal law has been affected by the [free-will] problem at every level: the definition of criminal offenses, the assessment of responsibility, including the practices we have adopted to reach such an assessment; and the way we deal with those found guilty, both in the formal sense of the institutions of punishment or treatment and in the informal sense of social views regarding the guilty."[12] And to give an example of how free will matters in civil law, note that the German Civil Code explicitly refers to free will as a precondition for legal competence, the capacity to amend one's rights, duties, and obligations—for instance, by entering into contracts, getting married, and so on.[13]

One can hardly begin to imagine how morality and the law would have to change if free will was found to be an illusion. In morality we would have to rethink our practices of holding one another responsible, and in the law we would have to reinterpret or perhaps even replace crucial notions such as voluntariness and criminal liability. Psychological studies further suggest that "disbelief in free will increases aggression and reduces helpfulness,"[14] makes people more likely to cheat,[15] and reduces support for retributive punishment—that is, punishment based on the idea that people deserve to be punished for wrongs they have culpably done.[16] Jean-Paul Sartre noted that "we are always ready to take refuge in a belief in determinism if this freedom weighs upon us or if we need an excuse."[17]

Of course, there are also some—such as the philosopher Derk Pereboom and the neuroscientist Sam Harris—who argue that embracing the nonexistence of free will does not mean that we have to abandon morality and law. Rather, "living without free will" could have positive aspects as well.[18] Critics of retributive punishment, for instance, might welcome scientific developments that challenge our conventional notions of responsibility and desert, and they might therefore find free-will scepticism congenial. As Pereboom reminds us, "Stoics maintained that we can always prevent or eradicate attitudes like grief and anger . . . with the aid of a determinist conviction."[19]

But even if one generally prefers a nonretributivist approach to punishment—one based on considerations other than retribution, for instance focusing more on preventing offences and on educating and rehabilitating offenders—it remains undeniable that the idea of free will and responsible agency is central to many aspects of human society.[20] It is therefore fair to say that giving up our belief in free will would require a significant revision in our understanding of the human condition.[21]

In short, free will, according to common sense, is a central human capacity, no less central than the capacities for thought and language. The challenge for science and philosophy is to clarify whether we really have this capacity and, if so, how it fits into the rest of our scientific worldview.[22]

Free Will versus Social Freedom

Before discussing the requirements of free will in more detail, I want to distinguish free will from some other notions of freedom. Free will must not be confused with social, political, or economic freedom. It is worth briefly explaining what those other kinds of freedom are. They all have to do with the constraints and opportunities we face in our social, political, or economic environments, and they are something that most of us care about a lot. For brevity, I shall simply speak of "social freedom." When Freedom House, an influential nongovernmental organization, rates different countries in terms of their freedom, what it is concerned with is social freedom, not free will in the psychological or agential sense discussed in this book. The claim that North Korea offers very little social freedom is not a claim about the psychological capacities of North Koreans *as human agents*. Their psychological capacities are the same as those of everyone else. Rather, the claim about North Koreans' lack of social freedom is a claim about the severe constraints they face in their social, political, and economic environment, and about the limited opportunities they have.[23]

How exactly to define freedom in this social sense is a contested matter, and proponents of different political viewpoints argue vigorously about it. But few political thinkers would deny that social freedom is valuable; they only quarrel about what the right definition of social freedom is.[24] Some, for example, think that social freedom merely requires the absence of external constraints, while others think that it requires substantive opportunities as well. For free-market liberals, unimpeded market transactions are central to social freedom; for left-leaning welfare-state liberals, access to resources matters too.[25]

All of these issues are important, but they are not the topic of this book. The philosophical question of whether there is a genuine human capacity of free will arises independently of the social conditions in which people live. It concerns the capacities we have as humans, not as members of any particular society. Consider an analogy. The question of how to make sense of the human capacity of language arises independently of the details of any particular language that people speak. Similarly, if free will is among the human capacities, then it is reasonable to think that human beings have this capacity independently of the details of their social environment. Indeed, we may presume that even early humans had this capacity, well before modern questions about social, political, and economic freedom ever arose—and well before people even had a *concept* of freedom.

This does not make the question of free will irrelevant to the debate about social freedom. Arguably, this debate implicitly presupposes that there is a human capacity of free will. If people had no free will, it would be much less clear whether there could still be genuine social freedom and—even if there could be—what exactly its significance would be. Moreover, we may think that exercising free will is valuable in itself. It would then follow that social conditions that give people more room to exercise this capacity are preferable to ones that give them less, other things being equal. In short, if we have free will, and we value its exercise, this may give us further reason to value certain forms of social, political, and economic freedom as well.[26]

The Three Requirements of Free Will

I have described free will as a three-part capacity: the capacity to act intentionally, to choose between alternative possibilities, and to control one's actions.[27] I will now make this more precise. My claim is this: For someone or something—say, a person or an organism—to have free will, three requirements must be fulfilled. I will call them "intentional agency," "alternative possibilities," and "causal control." Let me go through them one by one.

> **Intentional agency:** Any bearer of free will is an intentional agent, whose intentions support the relevant actions.

Thus, free will is a feature of intentional agents—actors with goals and purposes—and not of nonagential entities. Human beings are paradigmatic examples of intentional agents, so they clearly meet this requirement. Indeed, they are paradigmatic bearers of free will, according to common sense. By contrast, sofas, bicycles, and tomatoes, as well as other inanimate objects, do not meet the agency requirement. They are not actors like you and me, and they do not even begin to qualify as bearers of free will. It makes no sense to ascribe either intentional agency or free will to such a nonagential entity. How could something be said to have "free will" if it cannot be said to have "a will" at all?

As the theologian and philosopher Jonathan Edwards wrote in his 1754 book *Freedom of Will,*

> That which has the Power of Volition or Choice, is the Man or the Soul. . . . And he that has the Liberty of doing according to his Will, is the Agent or Doer who is possessed of the Will. . . . To be free, is the Property of an Agent who is possessed of Powers and Faculties, as much as to be cunning, valiant, bountiful, or zealous. But these Qualities are the Properties of Men or Persons.[28]

Of course, all this raises the question of what exactly an "agent" is, and what it means for something to be an intentional action of that agent rather than a mere physical event or behaviour. I cannot settle these questions at the outset, but I will come back to them at various points in this book. In any case, the basic claim that there is no free will without intentional agency should be uncontroversial.

Let me move on to the second requirement for free will:

> **Alternative possibilities:** Any bearer of free will faces the choice, at least in relevant cases, between two or more alternative actions: each is a genuine possibility for the agent.

This is the famous "could have done otherwise" clause. The idea that someone's choices are not free unless the agent could do otherwise is a familiar one. My choice of coffee over tea this morning was free, in part because I could have chosen otherwise. Choosing tea, in this trivial example, was a genuine possibility for me. Likewise, my choice of profession—in the lucky context of a prosperous society—was free, in part because I could have chosen a different career; I could have done something other than becoming a professor. To be sure, my choice of career was constrained by my opportunities and abilities. I could not have become an artist or an athlete, to give just two (of many) examples. And I would have probably failed if I had tried to become a financial trader, because I would not have had the right personality and skills for such a job. That said, most people—at least in prosperous societies, though unfortunately not everywhere—are not strictly confined to a single possible trajectory for their lives. It is because we can in principle choose among several options that we have alternative possibilities.

The possibility of doing otherwise is particularly salient in cases of wrongful harm. A murderer is responsible for his action, in part because he could have done otherwise, or so we commonly think.[29] Refraining from committing the act of murder was possible for him. Indeed, he *should* have done otherwise. And if it was truly impossible for him to act otherwise—for instance, because of a psychological compulsion—then he would be "not guilty by reason of insanity," as it is called in some jurisdictions. He would then not be a case for the criminal justice system, but would instead be given medical treatment.

Yet it is worth pausing for a moment and reflecting on whether having alternative possibilities is really necessary for free will. Some philosophers have challenged the view that it is. Daniel Dennett, for instance, recalls the story of Martin Luther, the sixteenth-century German monk who became a church reformer.[30] Luther provoked the Roman Catholic Church by criticizing some of its doctrines and practices at the time, such as its sale of indulgences. When he was summoned to the Diet of Worms in 1521 and was asked to renounce his criticism of the Church, Luther stood by his views and reportedly said, "Here I stand; I can do no other."

Does this mean Luther was denying that he had free will? If having alternative possibilities is necessary for free will, it would seem that he was. But surely this would be the wrong conclusion. Luther was not denying that he had free will; he was taking ownership of his actions, asserting that they were a consequence of who he was. But if taking ownership of one's actions is enough for free will, then it seems that alternative possibilities are not needed; being an intentional agent who endorses one's actions is the key requirement.

However, the story of Luther need not be interpreted in this way. What Luther was saying, on a more plausible interpretation, is not that it was literally impossible for him to do otherwise. Rather, he was not able to do otherwise *without sacrificing his integrity.* Had he done otherwise, which was within his abilities, he would have betrayed his values and probably regretted his choice. It is not that he did not have alternative possibilities; he just did not endorse them.[31]

For this reason, the Luther story is no counterexample to the claim that free will requires alternative possibilities. Rather, the story is fully consistent with the claim that both of the requirements I have introduced so far—intentional agency and alternative possibilities—are needed for free will. And so I will proceed on the assumption that not requiring alternative possibilities would be to water down free will.

Let me turn to the third requirement:

> **Causal control:** The relevant actions of any bearer of free will are caused, not merely by some nonintentional physical processes,

> but by the appropriate mental states, and specifically the intentions behind those actions.

For an action to be genuinely ours, we must have causal control over it.[32] Events over which we have no control are not up to us; they do not fall under the umbrella of our own free will. Furthermore, having causal control over our actions means not just that certain physical processes in the brain or body cause them. We would not say that we are in control of our reflexes, even though they are caused by physical processes in the body—for instance, when a knee reflex causes an involuntary leg movement, the kind of thing that happens when a doctor strikes your knee with a little rubber hammer.

Rather, for an action to be due to our own free will, it must be caused by our intentions. It must be true that we did the action *because we intended to do it.*[33] It is not enough for the action to be purely the result of some nonintentional (and subconscious) process over which we have no control. Just as bodily reflexes do not stem from our own free will, so we are not in control of our digestion, although it is regulated by our nervous system. In fact, we would not describe reflexes or digestion as actions at all; they are just bodily processes or physical movements.

In fact, even if a putative action is *in line with our intentions,* but it was not *caused by those intentions,* or it was caused in the wrong way, we would not say that it stems from our own free will. The philosopher Donald Davidson famously tells the following story:

> A climber might want to rid himself of the weight and danger of holding another man on a rope, and he might know that by loosening his hold on the rope he could rid himself of the weight and danger. This belief and want might so unnerve him as to cause him to loosen his hold, and yet it might be the case that he never *chose* to loosen his hold, nor did he do it intentionally.[34]

As Davidson observes, the climber's mental state did plausibly cause him to loosen his hold, and yet the causal chain was not the right one. The act in question was not properly under the control of his intentions. For this reason, we would be reluctant to describe what happened as an act of the climber's free will. Instead, we would say that the climber lost his nerves; he panicked and thereby caused an accident. In sum, causal control seems necessary for free will, over and above intentional agency and alternative possibilities.

Of course, I have not yet made the three requirements fully precise, but have left their interpretation somewhat open, thereby relying on the reader's pretheoretic intuitions. Clearly, there are several questions on which I must say more, and I will do so in subsequent chapters. In particular:

- What is an intentional agent?
- When is an action possible for an agent?
- What does it mean to say that an action is caused by a person's intentions?

For now, my claim is simply that there is a way of understanding the present three requirements such that they characterize free will in the conventional sense. That is, free will, in the broadly "libertarian" sense in which people commonly understand it, requires intentional agency, alternative possibilities, and causal control, suitably interpreted. I hope that the unbiased reader is more or less on board with this claim. Of course, I have not yet defended the distinct claim that we truly have free will in this sense. The three requirements are only meant to capture *what it would take to have free will, conventionally understood.*

Furthermore, many discussions of free will that we encounter in science and philosophy can be framed in terms of the three requirements.[35] In particular, some contributions to the debate can be viewed as attempts to show that some of the requirements *cannot* be met. Examples are the well-known arguments to the effect that determinism rules out alternative possibilities, or that neuroscience shows that we do not control our actions. I will discuss some of those later. Other contributions can be viewed as attempts to show that some of the requirements *can* be met. As Dennett and others have suggested, someone like Luther can certainly qualify as the *intentional agent* of his actions, whether or not these actions are predetermined. And a third set of contributions can be viewed as attempts to show that a subset of the requirements is enough to characterize free will, while the remaining requirements are not needed. Accounts that define free will exclusively in terms of *being the author of one's actions,* while setting aside alternative possibilities, are examples of this, by suggesting that what matters for free will is intentional agency alone.

However, I will here take a "maximalist" view and assume that all three requirements, suitably fine-tuned, are required for free will: I will take them to be jointly necessary and sufficient for free will. If I am able to show that we truly have free will in this strong sense, then this should hopefully vindicate the free-will presuppositions of most standard notions of agency, deliberation, and moral responsibility. By contrast, if we were to define free will in a watered-down manner, we would probably have to adjust our understanding of agency, deliberation, and responsibility as well. Otherwise our weakened notion of free will might not be able to play the conceptual role that free will is commonly assumed to play—for instance, as a condition for responsibility. Thus the question of whether we have free will in accordance with all of the three requirements is the most interesting and challenging one.

Free Will as a Capacity versus Its Exercise

As I have noted, my primary interest in this book is in free will as a capacity of an agent. Thus, *having free will* is a property of a person or some other entity or organism that qualifies as an agent. It is not tied to a particular action. At the same time, it makes perfect sense to ask, for each action in question, whether that action stems from the agent's free will. Was that action free?

I will say that an action is "freely performed" if and only if

- it is intentional—that is, appropriately supported by the agent's intentions;
- it is possible for the agent to do otherwise; and
- the action is under the agent's causal control.

Clearly the question of whether a particular action has these properties is distinct from the question of whether the agent in question possesses free will as a general capacity. A sleepwalker may well have free will under normal circumstances, and yet do erratic things during the night that have nothing to do with this capacity. Similarly, a drunken person may no longer be in a position to exercise her free will once intoxicated, even though she has freely chosen to drink in the first place.

If we wish to establish whether someone can be held responsible for something he or she did, we need to know not only whether the person has the capacity of free will in general but also whether what he or she did resulted from its exercise. Specifically, we need to know whether what the person did was freely performed, as characterized by the three bullet points above. Was it an intentional action? Could the person have done otherwise? Was the person in control? Or, if what the person did was not freely performed, we need to know whether the person's free will was at least implicated in the run-up to it: Was there a free decision to get drunk in the first place, for instance? Of course, moral responsibility might well require more than that (which is itself a difficult philosophical issue), but I do take the presence of free will somewhere along the relevant chain of events to be a *necessary condition* for a salient form of moral responsibility.

Even though it may be hard to say precisely *how much free will* is needed for moral responsibility, the basic idea should be clear. For example, the sleepwalker's erratic nighttime movement is not a free action by itself; nor is his free will relevantly implicated in the run-up to it. So, he cannot be held responsible for what he does while sleepwalking. Of course, if he knows that he is prone to sleepwalking and still ends up shooting someone during such an episode, we can hold him responsible for failing to take reasonable precautions. For instance, he ought not to keep a gun on his bedside table. Similarly, the drunken person may no longer act freely when heavily intoxicated, and so her drunk driving may not count as freely performed. Still, to the extent that her initial decision to drink was under her control and she did not take any precautions against drunk driving (such as leaving the car at home or giving the keys to someone else), her free will was implicated in the run-up to what she did, and she may thus be held responsible. As litigator Deborah C. England writes, "Intoxication is not an excuse for criminal conduct, but it may deprive an intoxicated person of the mental capacity to form the intent required by law to be convicted of certain crimes." However, she continues, "where a crime is defined by reckless conduct or negligence, intoxication will likely not be a defense, because it is foreseeable that alcohol will lead to reckless or negligent behavior."[36]

Needless to say, the issue of responsibility is a complicated one (and, at any rate, a separate topic), as is the issue of action attribution: What exactly does it mean for a given event or behaviour to count as the intentional action of a particular agent?[37] I will not be able to deal with these issues in detail here. The present remarks, however, should suffice to highlight the distinction between free will as a capacity and its exercise in a particular instance.

Free Will as a Matter of Degree

Finally, it is worth noting that we need not think of free will as an all-or-nothing matter. We may also recognize partial instances of free will—either in the case of an agent or in the case of a specific action—where only one or two, but not all three, requirements are satisfied, or where all three are satisfied, but only to a limited degree. Perhaps the agent in question has only a restricted range of agential capacities, for example. Or his or her causal control over a particular action is somehow compromised, though not completely absent.

How do we handle such borderline cases? Where do we draw the line between someone who has free will and someone who does not? And similarly, where do we draw the line between actions that count as freely performed, and ones that do not? In principle, once we make the three requirements for free will fully precise, we should be able to adjudicate every case by identifying which of the three requirements are met, and which are not. But, in practice, there may be different ways of "precisifying" these requirements: we can interpret them more stringently, or less stringently.

In light of this, the boundary between *free will* and *no free will,* and the boundary between actions that are freely performed and ones that are not, may be vague. When we try to classify agents and their actions in terms of "freedom," there may be clear cases on which we can readily agree, but there may also be contested cases, where different interpretations of our three requirements for free will yield different verdicts. A stringent interpretation of these requirements may then imply that there is no free will in a borderline case, while a less stringent interpretation may imply that there is. Different interpretations may be useful for different purposes. We might also simply refrain from drawing a sharp boundary and acknowledge that free will can be a matter of degree.

Many concepts have vague borderline cases. Think of *bald, tall,* and *rich,* for example. Sometimes it is hard to say whether someone is truly bald, truly tall, or truly rich; different interpretations of these concepts suggest different answers. Still, the lack of sharp boundaries does not undermine the existence of paradigmatic instances of each of these concepts. Lord Voldemort in the Harry Potter movies is truly bald; some basketball players are truly tall; and Bill Gates is truly rich.

Likewise, my modest proposal is that, under a fairly commonsensical interpretation, the three requirements I have introduced characterize free will in its paradigmatic form, and they also give us useful guidance on the conditions under which an action may count as freely performed. All this is compatible with recognizing that there may be different ways of drawing the line in borderline cases, and that sometimes the satisfaction of the requirements can be a matter of degree.

This completes my initial sketch of what free will requires. To sum up, I have argued that free will, in its paradigmatic form, is a three-part capacity: it requires intentional agency, alternative possibilities, and causal control. Furthermore, a specific action is freely performed if it is intentional, the agent could have done otherwise, and the action is under the agent's causal control. I will now turn to the question of whether we truly have free will in the sense introduced.

2

Three Challenges

I have argued that any bearer of free will—say, a person or an organism—must be an intentional agent; this agent must be capable of choosing among alternative possibilities; and he or she (or it) must have causal control over the resulting actions. But can we truly have free will in this sense? How does free will fit into a scientific worldview?

There are at least three major challenges for free will. I will call them the "challenge from radical materialism," the "challenge from determinism," and the "challenge from epiphenomenalism." They correspond, roughly, to the three requirements for free will that I have introduced. The aim of this chapter is to explain each of these challenges. I will begin with the first challenge, which targets the requirement of intentional agency.

The Challenge from Radical Materialism

This challenge can be summarized in terms of the following simple argument. It purports to deduce the nonexistence of free will from two premises:

Premise 1: Free will requires intentional agency.
Premise 2: Scientifically speaking, there is no such thing as intentional agency. This notion is a leftover from folk psychology and

premature versions of scientific psychology and will eventually be dispensed with in favour of a neuroscientific theory of human behaviour.

The first of these premises simply restates the first of our three requirements for free will. The second expresses the thesis of "radical materialism," which I will explain further in a moment. It should be evident that *if* we accept these two premises, *then* we must accept the following conclusion:

Conclusion: Scientifically speaking, there is no free will.

But should we accept the two premises? I have already argued for premise 1, the intentional-agency requirement for free will. But what about premise 2, the claim that intentional agency has no place in a scientific worldview? At first sight, it would seem absurd to reject the reality of intentional agency. After all, the idea that people are intentional agents, who form mental representations of the world and act in pursuit of goals and purposes, is central to our understanding of human behaviour. Commonsense psychology, which we use in everyday life to understand the people around us and to navigate the social world, is built around this idea. Consider the role that notions such as belief, desire, preference, and intention play in our understanding of our fellow humans. Even the most basic human interaction—say, a transaction in a shop—relies on certain assumptions about what other people think, want, expect, and intend. All these notions are bound up with the idea that people are intentional agents.

Moving beyond daily interactions, many theories in the social sciences also invoke the idea of intentional agency. Consumer theory in economics, to give just one example, represents market participants as rational utility maximizers. It would be virtually impossible for us to explain and predict human behaviour if we did not take human beings to be intentional agents, with beliefs and desires guiding their actions. Yet, some "materialist" thinkers, such as the neurophilosophers Patricia and Paul Churchland, have powerfully argued that this is not the best scientific perspective on human psychology.[1]

It is widely accepted that human cognition and behaviour are ultimately the result of complex biological and physical processes in the brain and body. At some fundamental level, the human organism is a biophysical machine. Many scientists and philosophers endorse a worldview known as "materialism" or "physicalism." This is the view, roughly speaking, that all phenomena in the world are either themselves physical phenomena or at least the result of physical phenomena: everything "supervenes" on the physical, in philosophical jargon. I already mentioned this idea in the Introduction. For example, chemical processes ultimately stem from physical processes; the laws of quantum mechanics underpin the way molecules are composed of atoms and the way they interact with one another. Biological processes stem from chemical and physical processes. Think of processes such as photosynthesis or the biochemistry of cells. Biology is a product of chemistry, which, in turn, is a product of physics. Finally, psychological processes stem from physical, chemical, and biological ones. They are implemented in a physical system: the human brain and body. This organism functions on the basis of a large and complex set of chemical reactions, and the brain and nervous system engage in information processing via electrical signals. If we accept a physicalist worldview, then we have no reason to think that the brain and body—and by implication, the human mind—stand outside the laws of physics. Rather, they are governed by the same fundamental laws that govern the rest of nature. That is the sense in which human beings are biophysical machines.

But now note that elsewhere in physical systems there appears to be no such thing as *intentionality,* where this involves a system's possession of mental representations of the world: beliefs, goals, and intentions. Rather, in physical systems, there are, at bottom, only physical mechanisms and law-like patterns. Intentionality does not seem to be a property of physical systems. It would then be odd to suppose that the human organism, itself a physical system, is an exception and somehow acquires the property of intentional agency.

Our tendency to ascribe intentions to people and some nonhuman animals, on this picture, is simply an evolutionarily advantageous trait, which allowed our ancestors to make sense of certain behavioural regularities and to predict them. Our ancestors had stumbled upon a useful fiction: that the world is inhabited by intentional agents. Hunters who interpreted their prey's behaviour as purposive—for instance, when an animal hides—were more successful than hunters who overlooked such goal-seeking behaviour in their prey. Furthermore, when our ancestors began to understand their fellow humans as intentional agents, ascribing mental states to one another, this gave them advantages in the way they could live together and coordinate their activities. It allowed them, among other things, to form increasingly complex societies. Recall again how important it is to understand the mental states of others when we go about our daily lives.

This explains why commonsense psychology evolved over the long history of the human species. However, so the argument continues, the usefulness of ascribing intentionality to others makes this no less fictional, and ascriptions of intentionality have no place in our best *scientific* understanding of the brain and behaviour. Even if we interpret one another as intentional agents for practical purposes, we can expect ascriptions of intentionality to disappear from science as our understanding of biology and psychology progresses.

Consider our ancestors' tendency to ascribe intentionality and purposes to many natural phenomena. People used to see intentional agency not only in the behaviour of humans and animals, but also in many other patterns in the world, from the weather to the plague. People thought that intentionality was ubiquitous. They believed in the existence of many intangible agents, such as ghosts, demons, and spirits, who could influence the physical world. Our ancestors might have viewed a storm or a flash of lightning as the manifestation of the intentions of a supernatural being, for example. In those cases, the human disposition to interpret the world in intentional terms, though otherwise useful, had some side effects: it led to overascriptions of intentionality.

A good illustration of this tendency to overascribe intentionality is given by a classic psychological study from the 1940s.[2] Participants were asked to describe and interpret the movement of simple geometrical figures in an animated film. Strikingly, they tended to do so in intentional terms, ascribing purposes and motives to these figures, even though they were just triangles and circles moving around a screen; they did not resemble human beings or animals at all. As Ilkka Pyysiäinen and others have observed, this tendency to make intentionality ascriptions wherever we see certain patterns might explain why the belief in supernatural agents is so pervasive in humans.[3] Humans would often rather ascribe intentionality to certain patterns in the natural world—whether it is a storm or a disease—than trace them to purely physical, nonintentional causes.

Step by step, science has shrunk the domain of "the intentional." We still take human beings to be intentional agents—and perhaps we do so too in the case of cats, dogs, and other complex animals—but we generally no longer believe in ghosts, demons, and spirits, and we prefer nonintentional explanations for most natural phenomena. As expressed by premise 2 in the little syllogism at the beginning of this section, the thesis of "radical materialism" assumes that it is only a matter of time until a neuroscientific understanding of the brain will drive out intentionality from science altogether. Currently, psychology is one of the last bastions of intentionality-laden discourse in science. Radical materialism postulates that the intentional idiom will eventually disappear from science completely. Philosophers also call this thesis "eliminative materialism."[4]

According to the proponents of this materialist thesis, such as Paul Churchland, commonsense psychology is a "stagnant or degenerating research program."[5] For a start, it leaves many important questions about the human brain and behaviour unanswered: Why is there sleep? How do we explain perceptual illusions? How does memory work?[6] Even more problematically, commonsense psychology invokes the outdated idea of an agent's mental representations of the world by means of inner states, such as beliefs and desires, which are hard to locate physically in the brain. As Churchland explains,

> A system of propositional attitudes [such as beliefs, desires, and intentions] . . . must inevitably fail to capture what is going on here, though it may reflect just enough superficial structure to sustain an alchemylike tradition among folk who lack any better theory. From the perspective of the newer theory [that is, neuroscience], . . . it is plain that there simply are no law-governed states of the kind [folk psychology] postulates. The real laws governing our internal activities are defined over different and much more complex kinematical states and configurations [such as neural states of the brain].[7]

In line with this, many recent advances in neuroscience offer explanations of human behaviour at a level of description very different from the intentional one.[8] As Helen Steward (a defender of free will who acknowledges the challenge) notes, we are now faced with an "increasingly large body of empirical evidence which seeks to root the explanation of an ever larger proportion of the things human agents do in sub-personal phenomena such as hormonal levels or neurally based predispositions."[9] For example, recent science suggests that "dopamine determines impulsive behavior,"[10] that genes affect our political orientation, such as whether we are liberal or conservative,[11] and that risk taking in teenagers can be traced to certain brain patterns ("greater connectivity between the amygdala . . . and the right middle frontal gyrus, left cingulate gyrus, left precuneus and right inferior parietal lobule").[12] Further, research in moral psychology suggests that whether people respond to moral dilemmas in more Kantian or more consequentialist ways depends on the extent to which those dilemmas trigger brain activity associated with emotional engagement.[13] So, a person's mode of moral reasoning may actually stem much more from subintentional brain processes than from high-level cognitive ones. And behavioural economics suggests that there is a fair amount of instinctive, subconscious, and "fast" decision making going on in humans, as opposed to "slow," conscious, and more explicitly deliberative reasoning, to use the terminology of the psychologist Daniel Kahneman, who distinguishes between "fast" and "slow" forms of thinking.[14]

If these developments exemplify the future of the behavioural and social sciences, then commonsense psychology may well await the same fate as some other "folk theories"—folk physics, folk biology, folk medicine, and so on: the kinds of informal, prescientific belief systems that humans habitually developed in relation to physical, biological, and medical issues in their everyday lives.[15] These folk theories were all useful up to a certain point, but were eventually superseded by very different and more scientific successor theories. Few of the concepts of the original folk theories in those other areas survived once we had a better understanding of the true fabrics of reality, whether in physics, biology, or medicine. And commonsense psychology, with its central concept of intentional agency, might be yet another one of those folk theories that are on their way out.

Now, one might say that the kind of radical materialism I have described is too extreme. A more reasonable thesis is that commonsense psychology is not strictly false, but that it offers insufficiently fundamental explanations of the phenomena at hand and that explanations of behaviour in terms of intentional agency will ultimately become redundant. On this picture, commonsense psychology will be *reduced* to something more fundamental. Again, neuroscience is the obvious candidate. It may not be strictly false to ascribe beliefs, desires, and intentions to people, but these are just shorthand descriptions for more fundamental neurophysiological properties of the underlying brains and bodies.

According to this less extreme form of materialism, which philosophers call "reductive materialism" (as opposed to "eliminative materialism"), the issue is not that there is no such thing as intentional agency at all. The issue is rather that talking about intentional agency is not strictly necessary, because such talk is fully translatable into talk about brain processes, in the same way in which statements about the temperature of water are fully translatable into statements about the mean kinetic energy of the water molecules. On such a view, saying that there is intentional agency over and above the underlying brain processes is entirely optional. Introducing the notion of intentional agency into our conceptual repertoire is not scientifically mandated, even though it is not strictly wrong. Again, the notion of free will stands on shaky ground. It relies on something we can reduce away: a leftover from an old-fashioned way of making sense of the world that can be replaced by something more fundamental. Talk of agency would then not be completely false, but still needlessly baroque.

To be sure, a lot more could be said about the forms of materialism I have described. But for now, I limit myself to acknowledging that they do put pressure on the picture of human beings as intentional agents that seems so central to our notion of free will. Let me move on to the second challenge, which targets our second requirement for free will: the requirement of alternative possibilities.

The Challenge from Determinism

This is probably the most widely discussed challenge for free will.[16] Science, especially in the form that goes back to the Enlightenment period, is often associated with the thesis that the world is "deterministic." This thesis says that the world is governed by mechanical laws of nature, according to which the past fully determines the future. Once the initial state of the world is given—say, its state at the beginning of time—all subsequent events unfold in an unalterable sequence, simply on the basis of the underlying laws of nature. The following argument summarizes the challenge that this poses for free will, again deriving a sceptical conclusion from two premises:

Premise 1: Free will requires alternative possibilities (among which an agent can choose).

Premise 2: If the world is deterministic—that is, the past fully determines the future—there are never any alternative possibilities: all actions are predetermined.

The first premise restates the second of our three requirements for free will. The second premise expresses an apparently direct implication of determinism. Together, the two premises entail the following conclusion:[17]

Conclusion: If the world is deterministic, there is no free will.

Note that, unlike the conclusion of the argument from radical materialism considered earlier, the present conclusion is a conditional one. It is not that there is no free will. Rather, it is that *if* the world is deterministic, *then* there is no free will.

To assess this argument I must say more about the idea of determinism. As already described, determinism is the thesis that, at any given time, the history of the world up to that time has only one possible continuation. We can think of this as a thesis about how constraining or "rigid" the laws of nature are. In a deterministic world, the laws of nature always permit just one possible future course of events at each point in time, given the past history. It could never happen that, when the past is given, the future may unfold in more than one possible way.

By "history" I mean not merely the kind of thing that historians are interested in and that we read about in history books: how Julius Caesar was assassinated, and when the Battle of Waterloo happened. Rather, the "history of the world" in the present technical sense is the sequence of all past states of the universe, from the beginning of time until the time of interest.[18] A "state of the universe," in turn, is a complete microphysical specification of everything in the world at a given point in time, including the location and momentum of each and every particle and the details of each and every physical force. "Determinism," then, is the thesis that, given the sequence of past states of the universe up to any point in time, there is only one way in which this sequence can continue: all subsequent states of the universe are necessary and inevitable consequences of the previous states.[19]

A deterministic universe is like a mechanical clockwork, where everything that happens is inexorably determined by what happened before. Isaac Newton's theory of physics represents the world like this, describing a set of laws under which the motion of any set of objects is fully determined by their initial states. For instance, given the initial states of all celestial bodies making up the solar system, Newton's laws specify what their movements will be forever after. The notion of the "clockwork universe" was the leading paradigm in the age of the Enlightenment. In a now famous passage, the French mathematician Pierre-Simon Laplace took the idea to its extreme:

> We may regard the present state of the universe as the effect of its past and the cause of its future. An intellect which at a certain moment would know all forces that set nature in motion, and all positions of all items of which nature is composed, if this intellect were also vast enough to submit these data to analysis, it would embrace in a single formula the movements of the greatest bodies of the universe and those of the tiniest atoms; for such an intellect nothing would be uncertain and the future just like the past would be present before its eyes.[20]

Of course, Laplace's demon, as this hypothetical intellect is often called, does not exist (as far as we know). And there is certainly no such intellect among mortals. But the point remains: in a deterministic universe, the future is predetermined by the past, even if there is no one around who is sufficiently smart and well informed to predict what is going to happen. Whatever happened at the beginning of time (say, during the first few moments after the Big Bang) would have been sufficient to necessitate everything that happened thereafter.

If our universe fits this picture, then your choice of beverage this morning, your choice of partner, your choice of career, and your choice of next year's holiday destination were all settled long before you ever thought about these choices. In fact, they were settled long before you were born. After all, you are part of the universe; and the physical building blocks that make up your brain and body are as much governed by the deterministic laws of nature as everything else is. It would then seem mistaken to think that you could have acted otherwise. There simply are no alternative possibilities. If this is right, your life is like a movie, the end of which is fixed before you take your seat in the cinema. It unfolds before your eyes, but there are never any forks in the road.

But is the world deterministic? Obviously, the present argument against free will does not apply if determinism is false. And determinism may be a remnant of an Enlightenment worldview that is no longer supported by current science.

Now, our current best theories of space, time, and large-scale physical systems—namely, Albert Einstein's special and general theories of relativity—still retain the determinism of Newtonian physics, despite superseding other aspects of Newton's theory. However, quantum mechanics, our best theory of physical systems at a microscopic level, appears to leave room for indeterminism. For instance, when a photon, a light particle, hits a semitransparent mirror (and we observe its trajectory), there is a 50 percent chance that it will be reflected, and a 50 percent chance that it will be transmitted. Nothing in the photon's prior history seems to determine which of these two possibilities will materialize. There appears to be a genuine fork in the road. Similar points could be made about other microphysical processes. Radioactive decay, for example, appears to be indeterministic. Even if we knew everything about the prior history of a particular uranium atom, we would still not be in a position to predict when exactly this atom will decay. The precise timing of its decay is left open by the atom's prior history.

Is this observation enough to defend free will against the challenge from determinism?[21] The answer is decidedly no. For a start, the interpretation of quantum mechanics itself is controversial. What everyone agrees on is that quantum mechanics supports some kind of "surface-level indeterminism": we cannot *predict* the photon's future trajectory when it hits the mirror, even given full information about its past. Similarly, we cannot *predict* when the uranium atom will decay, given complete information about the atom's past history. But this is where the agreement stops. While some scientists conclude that this surface-level indeterminism establishes that the world is fundamentally indeterministic (this view is supported by the so-called Copenhagen interpretation of quantum mechanics),[22] others disagree. Einstein famously said, "God does not play dice," thereby challenging not so much quantum mechanics itself but rather its indeterministic interpretation, which he found unsatisfactory.[23]

Interpretations of quantum mechanics that avoid this indeterminism include so-called hidden-variables interpretations, according to which the unpredictability of quantum systems stems from our ignorance of certain hidden variables—features of reality that are objective but not observable. These quietly determine how the system will evolve. In the example of the semitransparent mirror, the hidden variable will have put the photon either on track for reflection, or on track for transmission. It's just that, before running the experiment, we do not know what value the hidden variable has taken. For this reason, there is unpredictability here—we cannot predict what will happen to the photon—but crucially, there is no indeterminism.

The details of such hidden-variables interpretations are complicated. We know from mathematical physics that "local" hidden variables, which are tied to local processes such as individual photons, could not generally work (a result known as Bell's theorem). Instead, the hidden variables would have to be "global"—that is, attached to the physical system as a whole. These details notwithstanding, the mere possibility of interpreting quantum mechanics in a deterministic way shows that quantum mechanics by itself does not rule out determinism.

But even more importantly, our current best physical theories are not the final word on the laws of nature. Notoriously, quantum mechanics and the general theory of relativity, which are our current best theories of microscopic and macroscopic systems, are mutually incompatible, and there is no consensus on how to reconcile them—or rather, how to move beyond them in a way that avoids the conflict. It is therefore fair to say that the jury is still out on whether the grand unified theory of physics—if it is ever found—will represent the world as indeterministic.

It is also worth noting that merely showing that the world is indeterministic at some microphysical level is not enough to vindicate free will, even if we focus on the requirement of alternative possibilities alone. We would also have to show that these microphysical indeterminacies are amplified to a macroscopic level, where they can open up alternative courses of action for an agent. If the indeterminism was confined to the microphysical level and all indeterminacies somehow got washed out at the macroscopic level, this would not be enough for free will. The libertarian philosopher Robert Kane has suggested that the human brain has mechanisms in place by which quantum indeterminacies get amplified to a macroscopic level, but there is no consensus yet on the role, if any, that quantum-mechanical indeterminism might play in the brain's functioning.[24]

Finally, even if we could overcome all of these obstacles and show that fundamental physical indeterminacies have the effect of rendering multiple actions possible for an agent, there would still be the lingering worry that those indeterminacies introduce just randomness into human behaviour rather than free will. If indeterminism amounts to no more than randomness, then it is not clear how genuine *free choices* among the available options are possible, as opposed to a process of mere random picking. As Carl Hoefer, for example, notes,

> For reasons that Kant first realized, indeterminism at the microphysical level does not seem to help. The randomness, if any, in microscopic phenomena does not seem to "make room" for free will, but rather only replaces a sufficient physical cause with (at least in part) blind chance.[25]

In sum, the present challenge for free will is a formidable one, whether or not future physics will vindicate determinism.

The Challenge from Epiphenomenalism

The third challenge for free will arises even if the first two can be satisfactorily answered. It can again be summarized in terms of an argument with two premises:

Premise 1: Free will requires an agent's causal control over his or her actions; those actions must be caused not merely by nonintentional physical processes but by the relevant mental states, and specifically the agent's intentions.

Premise 2: Scientifically speaking, anything an agent does is wholly caused by nonintentional physical processes; the agent's intentions are, at most, byproducts of the underlying physical causes.

The first of these premises restates our third requirement for free will: the causal control requirement. The second premise expresses the thesis that an agent's behaviour has a purely physical cause: there is no such thing as "mental causation"—that is, causation by the agent's mental states. Those mental states are, at best, byproducts of the underlying physical processes—"epiphenomena"—which have no causal manifestations themselves. I will call this thesis "epiphenomenalism." Clearly, if we put those two premises together, we arrive at the following conclusion:

Conclusion: Scientifically speaking, there is no free will.

I have defended the first premise—the causal control requirement. But why should one accept the second—the epiphenomenalist thesis? Is it not manifestly true that when I intend to raise my hand and act on this intention, then the action is caused by my intention rather than by anything else? Perhaps there is a backstory to be told about how my brain and body implement this process at the neural level, but the action still seems under my control. It is my intention that is the causal source of the action, not anything else, or so we commonly assume.

However, there is an influential line of reasoning that puts pressure on our ordinary intuitions about mental causation. It is called the "causal exclusion argument" and has been prominently defended by the philosopher Jaegwon Kim, though it is arguably also implicit in much of the recent neuroscientific scepticism about free will.[26] For the moment, an informal summary will suffice. The argument proceeds by asserting two principles of causation that seem hard to deny from the perspective of a scientific worldview. And it then derives its negative conclusion—that there cannot be any mental causation—from those principles. Let's briefly run through them.

The first principle says that the physical world is "causally closed": anything that happens in the world must ultimately have a physical cause. There aren't any physically uncaused occurrences. To suppose that there are would be to accept a form of spooky supernaturalism. The avoidance of such supernaturalism seems integral to a scientific worldview.

The second principle prohibits the gratuitous overascription of causation. It says that once we have identified a fully sufficient cause for a given effect—a cause that fully accounts for the effect—then we must not attribute that effect also to some other allegedly fully sufficient cause, where that competing cause occurs *at the exact same time.* To hypothesize a second fully sufficient cause, over and above the first one, would be to postulate an implausible form of causal overdetermination. For instance, if a particular earthquake is fully sufficient to account for the resulting tsunami, then it would make no sense to attribute that tsunami also to another separate cause at the exact same time. If the earthquake is enough to account for the tsunami, then there is no further causal work for the purported second cause to do. To emphasize, we are not speaking about *earlier* causes here. There will of course have been *earlier* events which caused the earthquake itself—for instance, some previous tectonic movements—and which therefore indirectly contributed to causing the tsunami. The no-gratuitous-overdetermination principle only prohibits the attribution of rival causes that occur *simultaneously.* Kim also calls this the "causal exclusion principle."

If we accept the two principles—causal closure and causal exclusion—then we cannot avoid the conclusion that whenever someone performs some action, the cause must be some physical process in the person's brain and body rather than the person's intention. Since the action does not happen out of the blue, as a supernatural occurrence, it must have *some* cause. But then, postulating any cause other than a physical one—such as a mental cause—would breach one of the two principles. It would breach *either* the causal closure principle, if we were to insist that human actions can happen without any physical cause; *or* the causal exclusion principle, if we took the view that human actions have a mental cause *over and above* their fully sufficient physical cause.

To avoid both horns of this dilemma—one of which is supernaturalism, the other gratuitous causal overdetermination—we must conclude that anything an agent does is caused by some physical process in the person's brain and body. This, in turn, seems to leave little room for the agent's intentions to do any causal work. The upshot is epiphenomenalism: the agent's intentions and other mental states seem causally inert; they have no genuine causal manifestations. All the causal action takes place at the physical level.

Several neuroscientists have voiced support for epiphenomenalism, even though they do not explicitly invoke the philosophical argument that I have just summarized. The neuroscientist Sam Harris, for instance, writes,

> Did I consciously choose coffee over tea? No. The choice was made for me by events in my brain that I, as the conscious witness of my thoughts and actions, could not inspect or influence. . . . The intention to do one thing and not another does not originate in consciousness—rather, it appears in consciousness, as does any thought or impulse that might oppose it.[27]

Another neuroscientist, Michael Gazzaniga, tells the following story:

> If you were a Martian landing on Earth today and were gathering information how humans work, the idea of free will as commonly understood in folk psychology would not come up. The Martian would learn humans had learned about physics and chemistry and causation in the standard sense. They would be astonished to see the amount of information that has accumulated about how cells work, how brains work and would conclude, "OK, they are getting it. Just like cells are complex wonderful machines, so are brains. They work in cool ways even though there is this strong tug on them to think there is some little guy in their head calling the shots. There is not."[28]

So, the Martian scientist would conclude that anything human beings do is caused by physical processes in the brain. The intentions of the "little guy in the head" are at most an epiphenomenon: they do not "call the shots."

The views expressed in these quotations seem supported by a growing body of experimental work in psychology and neuroscience, much of which goes back to a classic study by Benjamin Libet and colleagues in the early 1980s.[29] Libet was interested in how a person's conscious intentions to perform an action relate to neuronal activity in the person's brain. He designed an experiment in which each subject—a college student—was asked to perform a simple action, such as pressing a button or moving a hand. The subject could perform the action at any time of his or her choice. Subjects were further asked to report the exact time at which they consciously formed the intention to perform the action. During the experiment they faced an easily readable clock, so they were able to monitor the time. This information allowed Libet to see how much time elapsed between a reported conscious decision and the action itself. Furthermore, throughout the experiment, Libet measured the electrical activity in each subject's brain via an EEG (electroencephalogram). In this way, he could observe the patterns of brain activity accompanying the subject's decision and action. Strikingly, the neuronal activity that led to the performance of the action—a neuronal readiness potential—could usually be detected a few hundred milliseconds *before* a subject became consciously aware of his or her intention to act. A natural conclusion, then, is that it is the subconscious brain activity that is causally responsible for a subject's action, not the conscious intention to act. The conscious intention is just an epiphenomenon. Commenting on Libet's experiments, the psychologist Daniel Wegner writes,

> It seems that conscious wanting is not the beginning of the process of making voluntary movement but rather is one of the events in a cascade that eventually yields such movement. The position of conscious will in the time line suggests perhaps that the experience of will is a link in a causal chain leading to action, but in fact it might not even be that. It might just be a loose end—one of those things, like the action, that is caused by prior brain and mental events.[30]

In a more recent study, John-Dylan Haynes and colleagues reported that brain-scan data could be used to predict, with better-than-chance accuracy, a subject's choice between two actions several seconds before the actual action took place, thereby replicating Libet's findings in an even more dramatic way.[31] A lot more could be said about how to interpret all of these experimental findings (and I will return to them in Chapter 5), but at least on their face they seem to provide evidence for epiphenomenalism: an agent's behaviour seems the result of physical causes, and the agent's intentions seem just a byproduct or a link in a longer causal chain.

It should be clear, then, that any attempt to defend free will is up against some major challenges. The claim that people truly have intentional agency would have to be defended against the challenge from radical materialism. The claim that they truly have alternative possibilities would have to be defended against the challenge from determinism. And the claim that they truly have causal control over their actions would have to be defended against the challenge from epiphenomenalism. In the following chapters, I will try to defend free will against each of these challenges.

3

In Defence of Intentional Agency

The first of our three requirements for free will is intentional agency: only systems that are intentional agents, such as human beings, can have free will. Any bearer of free will must be capable of acting intentionally, as opposed to producing mere nonintentional physical movements. Nonagents, such as rocks, tomatoes, and armchairs, do not even begin to qualify as bearers of free will. Of course, intentional agency alone is not enough for free will. The other two requirements—alternative possibilities and causal control—must be met as well; more on those in later chapters. Yet, even the intentional-agency requirement is demanding. Although we do not normally doubt that human beings are intentional agents, it is not obvious how intentional agency fits into a scientific worldview, as we have seen. The challenge from radical materialism suggests that this notion is a leftover from folk psychology and will eventually be dispensed with in our best neuroscientific theories of the brain and behaviour.

The aim of this chapter is to defend the commonsense view that human beings and other sufficiently complex animals are indeed intentional agents. I want to show that this is supported by our best understanding of the world. My claim is that intentional agency is a real phenomenon, not just an illusion or a useful fiction. Of course, to defend this view, I must explain where radical materialism goes wrong. My diagnosis, as will become clear, is that it overlooks two things:

1. The indispensability of the notion of intentional agency in the behavioural and social sciences.
2. The higher-level nature of intentional agency.

As I will argue, the notion of intentional agency is both practically and scientifically indispensable because there are many phenomena that we would not be able to make sense of without understanding humans and other complex animals as intentional agents. So, we would not be able to eliminate this notion from our discourse—even scientific discourse. And intentional agency is a higher-level phenomenon because it supervenes on physical phenomena but is not reducible to them. It is a genuinely "emergent" phenomenon. So, we would not be able to reduce intentional-agency talk to purely physical talk either.

My conclusion will be that human beings are truly intentional agents and thus meet the first requirement for free will. To make all this precise, let me begin by defining what an intentional agent is. Sometimes, for brevity, I will omit the adjective "intentional" and just use the terms "agency" or "agent" alone.[1]

What Is an Intentional Agent?

Informally speaking, an intentional agent is a system or entity that interacts with its environment in a goal-directed or purposive way. Human beings, but also other complex animals, are frequently cited examples. An intentional agent acquires and processes information about its environment, has goals and desires that it seeks to achieve, and acts in pursuit of these goals and desires, while taking into account its information. To give a simple example, a person's goal on a particular occasion may be to drink some coffee; the person acquires or retrieves some information about where to find coffee, and he or she acts by going there so as to achieve this goal.

More formally, we can define an intentional agent as an input-output system that meets at least three conditions:

1. It has representational states, which encode the system's "beliefs" about how things are—for example, the belief that coffee is available in a particular place.
2. It has motivational states, which encode the system's "desires" as to how it would like things to be, for example the desire to get some coffee.
3. It has a capacity to interact with its environment in pursuit of its desires, on the basis of its beliefs, for example by acting in such a way as to get some coffee.

Furthermore, a system of this kind is "instrumentally rational" to the extent that

- the system's beliefs and desires are internally consistent;
- the system's beliefs respond coherently to information it receives from its environment; and
- the system's actions are effectively guided by its desires, given its beliefs.

Obviously, the details of these requirements need to be spelt out further, but this is not my topic here. I will just mention briefly what sorts of things instrumental rationality demands. For example, someone who believes *both* that there is coffee available in a given place *and* that there isn't has inconsistent beliefs. This goes against the first of the three rationality requirements. Inconsistent beliefs and desires would not be properly action guiding. Second, consider someone who learns that the coffee has run out. If this person forms the belief that there is now *more* coffee available than before, he or she is responding incorrectly to the given information. This goes against the second requirement. An agent who doesn't meet this requirement would be insufficiently responsive to evidence; there would be a disconnect between the agent's beliefs and the environment. Finally, someone who desires to get some coffee and believes that coffee is available in place *A* rather than place *B* would be acting irrationally by going to *B* rather than to *A*. This is ruled out by the third requirement. Such a person would be unsuccessful on his or her own terms, by failing to pursue his or her own goals and purposes in light of the information he or she has.

On the other hand, someone who sets him- or herself a particular goal and then systematically pursues it, while impeccably taking all incoming information into account, is instrumentally rational. Whether the chosen goal is a worthy one is a separate issue; it is not adjudicated by instrumental rationality itself. Of course, the worthiness of an agent's goals is a very important issue, but it is not my topic here.

Finally, note that even if we focus on instrumental rationality alone, no intentional agent in the real world—certainly no *human* agent—is likely to be *fully* rational. Rather, instrumental rationality is best viewed as a standard of performance that intentional agents meet to greater or lesser degrees.

Intentional States and Intentional Actions

According to the definition I have given, any intentional agent has—at a minimum—certain representational and motivational states and a capacity to interact with the environment on the basis of those states. We can think of these states as key ingredients of the agent's "mind" or "psychology." Indeed, having a "mind" or "psychology" involves having some internal states that play certain roles in governing one's external interaction with the world.

Of course, an agent's mind and psychology may go beyond the simple belief and desire structure I have described. A human being has a great variety of psychological states above and beyond beliefs and desires. These include emotions, hopes and fears, and a large array of subconscious mental states, all of which play certain roles in the person's psychology and in his or her interactions with the world.[2]

Generally, intentional agents can come in many different shapes and sizes and can vary in their complexity and behaviour. A comparison between humans and other primates illustrates this point. Human beings are capable of very rich cognition and sophisticated planning, and both their mental processes and their interactions with the environment are facilitated by their use of language. Nonhuman primates such as chimpanzees do not have the same mental capacities and do not communicate via recursive language like humans, yet they still have substantial agential capacities. They use tools and engage in strategic behaviour, for example, and they pass the "mirror test." When they look at themselves in the mirror, they do not take this as an encounter with another animal and threaten their mirror image, but they recognize themselves and explore their own facial features. They seem capable of having mental representations of themselves.

Crucially, all intentional agents have one remarkable feature in common: the feature of "intentionality."[3] This feature is at the heart of any agent's goal-directedness. It is worth exploring this feature for a moment. A system's intentionality lies in the fact that the system has not just ordinary physical states, such as mass and momentum, but "intentional" ones. Intentional states are *about* something; they have a *meaningful content,* and they encode a certain *attitude* towards that content. For example, someone's belief that there is coffee available in a particular place has the content *that there is coffee available in that place,* and the attitude is a *representational* one: the content is represented as being true. Similarly, a person's desire to drink some coffee has the content *that the person drink some coffee,* and the attitude is a *motivational* one: the content is something the person would like to make true.[4]

Generally, intentional states can be defined as states that encode attitude-content pairs. The contents are typically propositions, and the attitudes encode the agent's relationship to those propositions, such as whether a given proposition is something the agent represents as true or is motivated to make true. Being "intentional" or "directed" in this sense is a semantic or logical property; it is not a physical or causal property. When an entity is in a certain physical state—for instance, it has a certain electrical charge or kinetic energy—this property can obviously have many *causal effects.* But such a physical state—construed solely as a physical state—does not have the kinds of *semantic or logical properties* distinctive of intentional states. Intentional states encode attitudes towards meaningful contents; physical states do not. Having meaningful contents, and encoding certain attitudes towards them, is what makes intentional states special. As a result, intentional states can be assessed in terms of instrumental rationality, and we can ask what actions they rationally support. Physical states do not lend themselves to such an assessment.

We can now see more clearly what makes something an "intentional action" as opposed to a mere physical movement or behaviour. Intentional actions are accounted for by certain underlying intentional states; in that sense, they are "rationalized" by those states. The attitudes and contents encoded by those states support those actions and render them instrumentally rational. This is very different from the case of mere physical movements or behaviours. These may have certain *causes,* but are not *rationalized* by them.

Suppose, for example, that I participate in a committee meeting at my university and vote on some proposal. My desire to cast a positive vote and my belief that I can do so by raising my arm make the resulting arm movement rational. The movement is an intentional action, specifically an act of voting for a particular proposal. My desire and belief *rationally support* that action; they do not merely cause it. By contrast, if my arm goes up accidentally due to a muscle cramp or a nervous reflex, this is not an intentional action; it is not supported or rationalized by any underlying intentional states. Even though there is some physical state of my body that causes the movement, such as a pattern of nervous activity, the relationship between that physical state and the action is a purely causal relationship, not an intentional one. The movement is not even a candidate for assessment as being "rational" or "irrational." It is subintentional and subagential: below the level of the organism's intentional agency.

A Test for Intentional Agency

It should now be clear what we mean by "intentional agency" and "intentional action." But how can we tell whether a given system is an intentional agent? Perhaps no system or organism genuinely meets the conditions for intentional agency. Could we be mistaken in thinking that humans and other complex animals are intentional agents?

What we need is a test that tells us, for any given system, whether or not that system is an agent. To devise such a test, I begin with a slightly different question—namely, how can we best *explain* a given system's behaviour? Obviously, different systems require different explanatory strategies. If we wish to explain the movements of the planets around the sun, we need to employ classical mechanics. Isaac Newton's famous principles, together with a specification of the relevant initial conditions, allow us to explain and predict the movements of the planets with good accuracy. Or, for another example, if we wish to explain how the distribution of heat evolves in a closed container over time, we need to use the heat equation, a partial differential equation for which we can compute a solution at various future times. Again, this will allow us to explain and predict the phenomenon of interest. Similarly, though less reliably, we can explain and predict the weather by modelling it as a stochastic process. In none of those explanations, however, do we need to invoke the idea of intentional agency. The systems in question can all be explained in purely physical terms.

By contrast, there are also systems—most notably, human beings and other complex animals—whose behaviour lends itself to "intentional explanation": explanation in terms of goals and purposes. Specifically, we can explain and predict such a system's behaviour on the basis of what we know or hypothesize about its intentional states and what behavioural responses those states would support or rationalize. Intentional agents are systems towards which we can take an "intentional stance," as Daniel Dennett puts it. He summarizes this idea as follows:

> First you decide to treat the object whose behavior is to be predicted as a rational [intentional] agent; then you figure out what beliefs that agent ought to have, given its place in the world and its purpose. Then you figure out what desires it ought to have, on the same considerations, and finally you predict that this rational agent will act to

> further its goals in the light of its beliefs. A little practical reasoning from the chosen set of beliefs and desires will in many—but not all—instances yield a decision about what the agent ought to do; that is what you predict the agent *will* do.[5]

If the system is truly an agent, and our hypotheses about the system's intentional states are broadly correct, then this explanatory strategy will yield reliable predictions. Dennett contrasts the intentional stance with two other stances that we might take towards a given system: a "physical stance," which would be warranted in some of our earlier examples (the solar system, the heat diffusion process, and the weather); and a "design stance," which would be warranted in the case of systems that are designed to fulfil certain functions but which do not qualify as intentional agents (a telephone or a washing machine, for example).

One may now propose the following simple test for intentional agency: A system is an intentional agent if it can be successfully explained in intentional terms. Dennett suggests something along these lines: "Anything that is usefully and voluminously predictable from the intentional stance is, by definition, an *intentional system*."[6] Dennett's view, however, appears to turn agency into something that is in the eye of the beholder. For Dennett, to be an agent is, in effect, to be *interpretable* as an agent. This view is also called "interpretivism" about agency.

To see why interpretivism is not entirely satisfactory, consider the example of a thermostat, which regulates the heating. As Dennett acknowledges, even a device as simple as this can be interpreted as an intentional agent.[7] It has "beliefs" about the actual temperature, which it updates in response to the input from its sensors, and it has "desires" about a target temperature—say, twenty degrees Celsius. It then "acts rationally" by regulating the heating, switching it on whenever the actual temperature is too low and off whenever it is too high. In this way, the thermostat is "usefully and voluminously predictable from the intentional stance." So, the interpretivist view would seem to imply that the thermostat is an agent. But doesn't this stretch the notion of agency too far?

One response would be to bite the bullet and to accept that a thermostat is an agent, albeit a very rudimentary one. On this picture, intentional agency is a fairly ubiquitous phenomenon. As Dennett puts it,

> Like the lowly thermostat, as simple an [animal as a clam] can sustain a rudimentary intentional-stance interpretation; the clam has its behaviours, and they are rational, given its limited outlook on the world. We are not surprised to learn that trees that are able to sense the slow encroachment of green-reflecting rivals shift resources into growing taller faster, because that's the smart thing for a plant to do under those circumstances. Where on the downward slope to insensate thinghood does "real" believing and desiring stop and mere "as if" believing and desiring take over? According to intentional-systems theory, this demand for a bright line is ill-motivated.[8]

Although I can see the appeal of this line of reasoning, my preferred approach to thermostats, clams, and trees is slightly different. Note that while it is *possible* to take an intentional stance towards such a simple system, this is presumably not *explanatorily necessary*.[9] We can easily explain the thermostat in terms of its mechanical design, and the clam or the tree in terms of its biological functioning, without making any ascription of intentional agency. So, it seems reasonable to amend our test for agency as follows:

> **The agency test:** To determine whether a system is an intentional agent, we must ask whether the system is *best* explained in intentional terms. For a system to pass this test, it must be the case that taking an intentional stance towards the system is not merely possible, but necessary or indispensable for certain explanatory purposes.

The revised test would allow us to distinguish between systems in relation to which the intentional stance is merely optional and systems in relation to which it is mandatory. The latter systems would then pass our test for agency; the former would not.

I want to emphasize that, by making use of this test, we are not committing ourselves to an interpretivist view about agency. According to interpretivism, to be an agent simply *is* to be interpretable as an agent. So, interpretability as an agent is what makes something an agent: it is *constitutive* of agency. The view I favour is different. Interpretability as an agent is not *constitutive* of agency, but merely *indicative*.[10] If a system is interpretable as an agent—especially if taking an intentional stance towards it is indispensable for certain explanatory purposes—then this is a good *indicator* that the system is an agent. But, as a matter of logic, being interpretable as an agent—even being *best* interpretable as an agent—is not the same as being an agent. It is just a strong symptom of agency. If something behaves like an agent and is most effectively explicable as an agent, this is pretty good evidence for agency. In the absence of any complicating factors, an inference to the best explanation would then support the conclusion that the system is in fact an agent.

The Indispensability of Agency Ascriptions

We have noted that systems ranging from telephones and washing machines to thermostats, clams, and trees do not meet our test for intentional agency: ascribing agency to them is not necessary in order to make sense of them. Radical materialism insists that humans and other mammals are no different. Although they are more complex than those simpler systems, and folk psychology treats them as agents, they are ultimately nothing more than elaborate biophysical machines. On this view, there is no place for intentional agency in science.

If radical materialism is right, then humans and other complex animals should fail our test for agency. Even if taking an intentional stance towards them is possible, it should also be possible, and in fact scientifically preferable, to explain their behaviour without ascribing intentional agency to them. In short, humans and other complex animals are not *best* explained in intentional terms. To see whether that claim is correct, let's begin by imagining you are a dog owner. Suppose your dog is used to being fed in the kitchen. Every day, at a particular time, you place a bowl of food on the kitchen floor. The dog, which has been running around in the garden, goes into the kitchen and eats the food. How do you explain why it goes to the kitchen, rather than to the living room, the corridor, the garden shed, or any other place?

The answer is easy. The dog is hungry and desires to eat. It believes that there is food in the kitchen. Perhaps it can even see you prepare the food, or smell it. And so, the rational response for the dog is to make its way to the kitchen rather than somewhere else. The dog exhibits goal-seeking behaviour. It is an intentional agent, though obviously not at the human level of sophistication.[11]

Now the proponents of radical materialism will say, Not so fast. The ascription of agency to the dog is a metaphor that may be useful for some everyday purposes, but it is not vindicated by science. Or, if we go with the less radical form of materialism—the "reductive" rather than "eliminative" one—it would seem that, even if it is not strictly false to suggest that the dog displays goal-seeking behaviour, it is not the best account. A more scientific explanation of the dog's behaviour, so the challenge goes, would describe the underlying processes in the dog's brain and body as it moves towards the kitchen; there would be no need to ascribe goals and purposes to it.

At this point, however, things get complicated. We would have to give a detailed account of all the relevant bodily and neurophysiological mechanisms inside the dog. What we conventionally describe as the dog's "getting hungry" would have to be redescribed in terms of complex physiological processes. And the dog's movement to the kitchen would have to be explained in terms of a dizzying number of neural firings in its brain, the transmission of a large volume of signals throughout its body via its nervous system, and the resulting activation of the relevant muscles.

The length and complexity of the physical details would be formidable. The dog's cerebral cortex alone has more than 500 million neurons, many of which will be implicated in the process, and its body consists of billions of cells.[12] The task of explaining, in physical rather than intentional terms, why the dog goes to the kitchen would be nothing short of mind-blowing, even if we do not insist on giving a full microphysical explanation and are content instead to explain the dog's behaviour at the level of its brain functioning. At most, we might be able to explain some *fragments* of the relevant process, such as what neural activity is triggered when light hits the dog's retina, or how different neurons in the dog's visual cortex fire in synchrony when they encode different features of the same perceived object.[13] Physically explaining the dog's behaviour in its entirety, however, would involve an informational and computational overload.

But let us suspend all disbelief and imagine that neuroscientists have somehow come up with a biophysical explanation of why the dog goes to the kitchen rather than somewhere else on the present occasion. This explanation would presumably enumerate all the details of the relevant neurophysiological processes in the dog's brain and body. Now imagine a slight change in circumstances. Today you decide to feed the dog in the garden shed rather than in the kitchen. The dog, while running around, sees you carry the usual bowl of food from the kitchen to the garden shed. Naturally, you predict that the dog will follow you into the garden shed, instead of going to the kitchen. Our original intentional explanation can easily accommodate this change. As before, the dog desires to eat. Based on what it sees, it now comes to believe that there is food in the garden shed, not in the kitchen; and so it responds by going to the garden shed. This is because the dog is capable of updating its beliefs—its representation of the environment—in light of new sensory information, and it rationally adjusts its actions accordingly. By understanding the dog as an intentional agent, we are not only able to explain its behaviour in the original case—its going to the kitchen for food—but we are also able to predict how it will respond in the new circumstances, when the food has been moved. We are thus able to explain the dog's behaviour in a flexible, yet parsimonious way.

Consider how the proponents of radical materialism would handle this change in circumstances. Let us assume, as we have hypothetically granted, that we have enumerated all the biological and physical details of the dog's behaviour in the original situation. Could we then predict the dog's behaviour in the new situation—with food in the garden shed rather than the kitchen—simply by looking at the original biophysical details of the dog's movement and adjusting them? It is hard to believe that we could. Of course, the dog's behaviour in the new situation—its going to the garden shed—will also have a complicated physical backstory. As before, there will be a dizzying number of neural firings in the dog's brain, a huge volume of signalling in its nervous system, and some resulting muscle activation. But the details will be subtly different from those in the earlier situation. The original, incredibly complex physical story will have to be replaced by another, equally complex one. It is entirely unclear whether the new story could be easily derived or extrapolated from the earlier one. In other words, even a full enumeration of the physical details of the dog's earlier movement to the kitchen may not be a good guide to what will happen in the new situation. And so it is questionable whether a biophysical explanation of the dog's behaviour would have much predictive power.

A good scientific explanation should not only describe the things that actually happen but also give us some indication as to what would happen in different circumstances. The biophysical explanation of the dog's behaviour overwhelms us with microscopic details without identifying what really lies behind the dog's movements—namely, its search for food, guided by its cognitive representation of the environment. By refusing to ascribe representational and motivational states to the dog, the proponents of radical materialism will have deprived themselves of the ability to identify the most salient regularities underlying the dog's behaviour and to make predictions on the basis of those regularities. They will have missed the dog's goal-seeking nature.[14]

Unsurprisingly, biologists in areas such as behavioural ecology use ideas from economics and the social sciences—especially decision theory and game theory—to explain and predict animal behaviour. In doing so, they recognize that animals behave in ways that involve optimization, strategic calculation, and planning: animals respond to incentives. Decision theory and game theory, which allow us to explain such behaviour, can be viewed as mathematical formalizations of the belief and desire model of agency. Saying, for instance, that someone—a person or an animal—behaves so as to *maximize expected utility* is a formal way of saying that he or she pursues his or her desires or goals according to his or her beliefs. Desires or goals are formally captured by the notion of "utility," and beliefs are formally captured by the reference to "expected" utility, where the "expectation," in turn, is defined relative to some "subjective probabilities." In this way, decision-theoretic and game-theoretic explanations can be seen as instances of intentional explanations.

The reason why biologists use such explanations is not that they reject the basic assumption of cell biology and biochemistry—namely, that life is the result of physical and chemical processes and ultimately governed by the physical laws of nature. Rather, the reason they employ (what are in effect) intentional explanations is that these are hard to avoid for some explanatory purposes. We would have great difficulty making sense of the rich patterns of animal behaviour—from nest building and foraging to hiding and cooperative hunting—if we did not recognize the goal-seeking nature of animals. The quest for adequate and predictively successful explanations often demands that we ascribe intentionality and sophisticated cognitive capacities to animals.

Now let us turn from the case of animals to that of humans. If we often cannot make sense of the behaviour of nonhuman animals without ascribing intentional agency to them, what hope is there of making sense of human behaviour without such agency ascriptions? There is one thing that practically all the human and social sciences have in common, from anthropology and psychology to economics, political science, and sociology: they all represent human beings as intentional agents.

To be sure, there is much disagreement about the precise nature of human agency, and social scientists quarrel about what the correct model of human psychology is. For example, many social scientists find the kinds of decision-theoretic or game-theoretic models that are traditionally used in economics too simplistic (and rightly so), and there is much work at the overlap of economics and psychology that seeks to improve upon those models.[15] Also, different social scientists disagree about the extent to which society is shaped by the agency of individual people as opposed to structural factors, such as institutions, culture, geography, and other constraints.[16] Yet the fundamental idea that human beings are intentional agents, who behave in a goal-seeking manner, is not in dispute. None of the established human and social sciences could get off the ground if we seriously denied that human beings are intentional agents.

This point should be evident enough, so that there is not much need to dwell on it further. But just to make it more vivid, ask yourself how you would answer the following questions if you could not draw on any hypotheses about people's beliefs, desires, and goals—that is, without ascriptions of intentional agency:

- Why do people turn out to vote?
- Why do people usually show up for work on weekdays, but stay at home on weekends or on days on which their businesses are closed?
- Why does the price for a good depend on supply and demand?
- Why do people give more money to charity when a major disaster has happened, such as an earthquake or a tsunami?
- Why might the presence of a surveillance camera reduce shoplifting in a supermarket?
- Why do legislators engage in logrolling?
- Why can you normally rely on other people's promises, especially if you know these people well and they strike you as trustworthy?

Obviously, some of these questions are harder to answer than others, but even the harder ones do not require rocket science. As long as you are willing to make some hypotheses about the relevant people's beliefs, goals, and other psychological states, you have a perfectly workable strategy at your disposal for explaining the phenomena in question. On the other hand, if you viewed people as nothing but biophysical machines, you would not even know where to begin.

In short, it is safe to conclude that ascriptions of intentional agency are indispensable for making sense of the social world, in everyday life and in the social sciences alike.[17] Of course, none of this is to say that we should not study the neural foundations of human psychology and behaviour. Figuring out how the biological brain gives rise to strikingly complex cognitive capacities is one of the major scientific challenges of our time. And there have been exciting neuroscientific advances in recent years. To give just a few examples, we now know more than before about how the brain manages to encode feature binding in perception: the attribution of different features to the same perceived object.[18] We also have a better understanding of the neural foundations of conscious awareness. This gives us insights into the mechanisms by which the brain physically implements conscious cognition, and it potentially opens up new treatments for various neurological conditions.[19] And psychologists and neuroeconomists are beginning to uncover some of the neural correlates of people's attitudes towards risk. High levels of the neurotransmitter dopamine in the brain have been associated with increased risk taking, for instance.[20] Still, given what I have argued, insights in neuroscience are best understood as *complementing* intentional explanations in the human and social sciences, not as *replacing* them. And so, I think, human beings and other complex animals pass the test for intentional agency. For many explanatory purposes, they are best understood in intentional terms. Taking an intentional stance towards them is not merely optional but mandatory.

The Higher-Level Nature of Intentional Agency

I have argued that the intentional agency of human beings and other complex animals is a real phenomenon, not a leftover from a prescientific age. Nonetheless, we might have the suspicion that this phenomenon is somehow "reducible" to underlying physical phenomena. Perhaps reductive materialism, the slightly less radical form of materialism, still has a point. Recall that, according to reductive materialism, it is not strictly false to say that humans and other animals are agents, but it is just a shorthand. Any talk of intentional agents can, in principle, be translated into a more scientific account of brain processes and bodily physiology. So, what can be said in response to this claim?

The first thing to say is that, even if this reducibility claim were true, it would not by itself undermine intentional agency as a real phenomenon. To give an analogy: just because any talk of "temperature" can be translated into a microlevel talk of "mean molecular kinetic energy," it does not follow that temperature is unreal. Temperature is a real phenomenon, notwithstanding its reducibility to something more fundamental. At most, the notion of temperature might be redundant for some scientific purposes (namely, in case our more fundamental theory can explain the physical world without mentioning temperature). Similarly, the reducibility of intentional agency to something more fundamental would not imply that it is unreal. But might the notion be redundant? We have already seen that our ascription of agency to humans and other animals is not at all redundant but explanatorily indispensable. We would have a hard time making sense of the living world—both in ordinary life and in the behavioural and social sciences—without interpreting people and other animals as intentional agents.

Second, and more importantly, it is questionable whether any talk of intentional agents is in fact reducible to a physical or biological account of processes in the brain and body. To explain this point, I must first say more about what "reducibility" would mean. The example of temperature offers a good illustration. The higher-level property of "the temperature being such-and-such" is reducible to something more fundamental because there exists a microphysical property that is equivalent to the higher-level property—namely, "having such-and-such mean molecular kinetic energy." Of course, "equivalence" here does not mean *equivalence in linguistic description.* Clearly the notions of "temperature" and "mean molecular kinetic energy" are described in very different terms. Rather, "equivalence" means two things:

1. Equivalence in satisfaction conditions: The lower-level and higher-level properties are necessarily co-occurrent. There couldn't be a certain temperature without a certain mean molecular kinetic energy and vice versa.

2. Substitutability for scientific explanatory purposes: The lower-level property can serve as a substitute for the higher-level property in scientific discourse. Whenever we employ the notion of temperature in some scientific explanation, we could, in principle, also use the notion of mean molecular kinetic energy.

More generally, we say that a higher-level property is "reducible" to some lower-level property if there is a lower-level property that is equivalent, in the relevant sense, to that higher-level property. Slightly more formally expressed, if $\mathbb{P}$ (note the outline font) is the higher-level property in question, there must exist some lower-level property P (note the normal font) such that $\mathbb{P}$ is equivalent to P. More generally, P could also stand for some combination of properties, such as P_1 *or* P_2 *or . . . or* P_k. The equivalence of the higher- and lower-level properties, in turn, requires that

1. $\mathbb{P}$ is instantiated by a system if and only if P is instantiated; and
2. for scientific explanatory purposes, P can serve as a substitute for $\mathbb{P}$.

In the temperature example, $\mathbb{P}$ would be "having such-and-such temperature" and P would be "having such-and-such mean molecular kinetic energy."[21]

If all properties to which we refer in some higher-level domain of discourse can be reduced to corresponding lower-level properties, we can, in principle, arrive at a translation between the two domains of discourse. We can then translate every higher-level statement into a lower-level one. Perhaps the resulting lower-level statement will be more complicated than the higher-level statement, so that our higher-level discourse remains useful for practical purposes. But the higher-level domain of discourse could in principle be "reduced away." It is nothing more than a useful shorthand for something that can equally be expressed at the lower level, at least in principle.

In the case of temperature, this picture may well be true. Our talk of temperature is nothing more than a useful shorthand for something we can equally express at a more fundamental level. If reductive materialism is right, then the same will be true for intentional agency. Our talk of agency may be a useful shorthand for a more complicated talk of neurophysiological processes, but the former will still be reducible to the latter. I will now explain, however, why this reducibility claim is implausible in the case of agency. There are at least two reasons for this; one is conceptual, and the other is combinatorial. Let me begin with the conceptual reason.

Conceptually, the obstacle to reduction lies in a feature of agents that I have already emphasized: their intentionality. Recall that a system is intentional if some of its states, such as its belief-and-desire states, are directed towards something: they encode an *attitude* towards some *meaningful content.* In the case of beliefs, the attitude is representational, and the content is a proposition that is represented as true. In the case of goals or desires, the attitude is motivational, and the content is a proposition that the agent would like to make true.

As it is often put, intentionality has to do with "the directedness, aboutness, or reference of mental states—the fact that, for example, you think *of* or *about* something."[22] Although "aboutness" (a term that goes back at least to John Searle) is not the nicest word, it captures what is special about intentional states: they are about something.[23] So, let us say that whenever an agent has a particular intentional property, such as "believing that Washington, DC, is the capital of the United States," "desiring a better world," or "intending to go swimming," this has the feature of "aboutness": it is about something. It is this aboutness that allows those intentional properties to play certain roles in rationalizing an agent's actions.

As noted earlier, my beliefs and desires do not merely *cause* my actions, but they render them *instrumentally rational.* By contrast, physical properties, such as "having a certain number of protons," "having a certain electrical charge," or "having a certain weight," lack any aboutness: they are not about anything. Physical properties can stand in *causal* relations with other properties, but, conceptually speaking, they do not stand in the *rational* or *semantic* relations akin to beliefs and desires.

We are now in a position to see why—from a conceptual perspective—intentional properties, such as believing, desiring, and intending, are not plausibly reducible to physical properties. Take any intentional property of an agent, such as "believing that Washington is the capital of the United States." To reduce this property to the physical level, we would have to identify some physical or neurophysiological property (or a combination of such properties) such that the given intentional property is equivalent to that lower-level property (or property combination). As noted earlier, equivalence requires two things: first, necessary co-occurrence of the higher- and lower-level properties and, second, substitutability for scientific explanatory purposes. Let's suppose, for the sake of argument, that the first requirement can be met. This means that we can find some neurophysiological property—say, a complicated pattern of neuronal firings—such that an agent's brain has that neurophysiological property if and only if the agent believes that Washington is the capital of the United States. (I will later raise some doubts about whether we can really find such a neurophysiological property.) This neurophysiological property would necessarily co-occur with the given belief property and would, in that sense, qualify as its lower-level counterpart. But what about substitutability for scientific explanatory purposes?

Remember that the intentional property of "believing that Washington is the capital of the United States" has the feature of "aboutness": it encodes a meaningful content. By contrast, the identified neurophysiological property lacks this feature. Why? By being a purely physical or biological property, it is, conceptually speaking, below the level of intentionality. It stands in various *causal* relations to other physical properties, but it stands in no *semantic* or *logical* relations, such as relations of rational coherence with other intentional properties or relations of reference to objects such as Washington or the United States. Now, even if the particular neurophysiological property always co-occurs with the given intentional property—that is, the physical brain has the neurophysiological property if and only if the agent's mind has the relevant belief—the two properties are explanatorily distinct: there are some explanatory roles that the intentional property can play and the neurophysiological property can't.

Because the intentional property features in semantic or logical relations, it can serve as an ingredient in *intentional* explanations of an agent's behaviour—for instance, by rationalizing certain actions. The neurophysiological property, by contrast, can serve at best as an ingredient in *causal* explanations. And so, the two properties are not substitutable for scientific explanatory purposes. The lower-level property is, by its nature, unsuitable for intentional explanations, and, as I have argued, these are often indispensable. For this reason, the attempt to reduce intentional properties to physical ones hits a conceptual roadblock. To put it bluntly, thinking and intending are properties of the mind, not of the brain. The brain is the site of physical processes. It is only the mind that is the site of thought.[24]

Let me now turn to the second reason why a reduction of intentional agency to the physical level may not be feasible: the combinatorial reason. Combinatorially, the obstacle to reduction comes from another widely assumed feature of intentional properties:

> **The multiple realizability of intentional properties:** The same intentional property may be instantiated by a variety of different physical properties or combinations of physical properties. There may be little, in physical terms, that these different possible "realizers" of the intentional property have in common, apart from the fact that they all instantiate the same intentional property.

Generally, whenever a higher-level property is instantiated by means of some configuration of lower-level properties, we call the latter the "lower-level realizer" of the given higher-level property. A higher-level property that is "multiply realizable" admits more than one possible lower-level realizer. If intentional properties are multiply realizable, then there may not be a single unified physical property P within the brain and body or even an easily specifiable combination of such properties, like P_1 *or* P_2 *or* . . . *or* P_k, which is the exact physical counterpart of the given intentional property $\mathbb{P}$.

To explain this, let me begin with an example of multiple realizability from a different domain—distinct from the context of agency and free will. Consider the property of "having one hundred dollars," which belongs to the social rather than the psychological domain. Think of all the different ways in which this property could be instantiated.[25] One could have a particular printed piece of paper in one's wallet (namely, a one-hundred-dollar note) or a collection of coins (for instance, four hundred quarters). Or one could have a different configuration of pieces of paper and / or coins (such as five twenty-dollar notes, or nineteen five-dollar notes and twenty quarters); and those notes or coins could be located in a different place to which one has access, where "access" could itself mean a variety of things. Alternatively, one could have a bank account with one hundred dollars in it, which can also be instantiated in different ways. Nowadays, it will be stored in the bank's electronic database, which, in turn, is distributed across a computational cloud. In the old days, it would have been recorded in a traditional book and other written records. Even more abstractly, having one hundred dollars could consist in standing in certain relationships with other people, which come with particular debt obligations. Those relationships could themselves be instantiated in many different ways.

If $\mathbb{P}$ stands for the higher-level property of "having one hundred dollars," it may be outright impossible to identify a single physical property P that is instantiated *if and only if* the higher-level property $\mathbb{P}$ is instantiated. Of course, one might try to identify one distinct physical property for each possible instance of "having one hundred dollars": one physical property for the case in which you have certain banknotes in your wallet; another for the case in which you have coins; a third for the case of a traditional, paper-based bank account; a fourth for an electronic bank account, and so on. In reality, however, each of these cases will itself have innumerably many subcases, which are all physically distinct from each other. A little reflection suggests that the attempt to identify all the possible physical patterns that could instantiate the higher-level property of "having one hundred dollars" is like the task of Hercules fighting the Hydra: for each head that is chopped off, two new heads will appear. Each time we identify a possible lower-level counterpart of "having one hundred dollars"—such as "having one hundred dollars in a bank account"—it tends to bifurcate into two or more subcases with diverse features. In the end, there may not be a manageable list of physical properties that exhausts all the different ways in which the given higher-level property could be realized. The higher-level property $\mathbb{P}$ may be equivalent, at most, to an *infinite* disjunction of physical properties, of the form

either P_1 *or* P_2 *or* P_3 *or . . .*

The only thing that the individual elements of this disjunction (P_1, P_2, P_3, and so on) have in common may be the fact that they are instances of the same higher-level property of "having one hundred dollars." The higher-level property then defies reduction.

As philosophers such as Hilary Putnam, Jerry Fodor, and others have argued, the attempt to reduce a mental, intentional property to an underlying physical property is prone to run into the same difficulties.[26] Consider again the intentional property of "believing that Washington is the capital of the United States." I have this property now. So do you, I assume, and all likely readers of this book. I also had this property when I was a child with a developing brain, taking geography lessons at school, but I didn't have it at preschool age. Likewise, millions or billions of people, young and old, with all their human differences, share this property now, or acquire it at some point, or had it in the past. The challenge for the proponent of a reduction is to pinpoint a single physical property or a readily specifiable disjunction of physical properties which all the actual and possible bearers of this belief—and no one else—have in common. Moreover, the physical property must be describable in purely physical or neurophysiological terms, without recourse to intentional language. It would be cheating to say that the physical property is "whatever property all brains share which encode the belief that Washington is the capital of the United States." If we tried to describe the target property like this, we would be using intentional language.

Now, there might not be a unique physical marker that is shared by all and only those brains that encode the belief in question. If there were a special "Washington neuron," for example, then we might be able to achieve the desired reduction. The neuron would be common to all and only those people who believe that Washington is the capital of the United States, and having the belief would be equivalent to having that neuron. But although there is a debate on whether individual thoughts—such as thoughts about Washington, or thoughts about your grandmother—are encoded by a small number of neurons or distributed across a larger neural network, it is doubtful whether we would ever be able to capture, in purely physical terms, what a person thinks.[27]

Even if the number of neurons encoding each person's beliefs about Washington were as few as eighteen thousand—a number that is sometimes quoted by neuroscientists[28]—this would not guarantee that it will ever be feasible to provide a purely physical or neurophysiological description of the relevant neural pattern: a description that does not use any intentional language. The description would have to be neither overspecific nor underspecific: it would have to correspond to the presence of the relevant belief and to nothing else. But even if we managed to pinpoint the relevant neural pattern in one person's brain—presumably by carefully monitoring the person's brain activity in response to various stimuli—it is still far from clear whether the *exact same* pattern would carry over to other people. Different brains are different, and even in the same person different neural configurations can sometimes play similar roles—a feature known as "neural plasticity." To make matters worse, there is no reason why the belief that Washington is the capital of the United States could not also be realized in a very different hardware, distinct from the ordinary biological brain—for instance, in the computer simulation of a neural network or in a radically different computational architecture.

Philosophers sometimes distinguish between "token reduction" and "type reduction." A "token reduction" of a given intentional property would be achieved if each *instance* of that property could be redescribed in physical terms, so that we would be able to point to the physical pattern that underpins *any particular instance* of the property, without necessarily carrying over to other instances. To go back to the earlier example: my having one hundred dollars *on a particular occasion* may be physically realized by the presence of certain banknotes in my pocket. But the same configuration of banknotes is not shared by other instances of the same property. A "type reduction" requires that we are able to point to a general property (or property combination) at the physical level such that, *in all possible cases,* the intentional property is instantiated if and only if the physical counterpart is. The problem of multiple realizability challenges the possibility of such a type reduction. The bottom line is that even though intentional properties may ultimately depend on underlying physical properties of the brain—as physicalists or materialists typically accept—this does not imply their reducibility.

It is worth considering an analogy. Think about the relationship between the individual pixels on a computer screen and the resulting image.[29] Clearly, if we fix the configuration of pixels—their locations and colours—this fixes the image, such as an image of the Mona Lisa. Once you arrange the pixels in a certain way, the image is an automatic consequence. This is what we mean by saying that the image "supervenes" on the pixels.[30] Similarly, we may argue that a person's intentional properties supervene on the physical properties of his or her brain and body. Once a person's brain is what it is, and all the ongoing patterns of neural activity are what they are, this presumably gives rise to the person's intentional states. The person's beliefs, desires, and goals could be no different unless there were some manifestation of that difference—however subtle—in the underlying neural activity.

But although the image supervenes on the pixels, the image may have certain holistic properties that cannot be found at the level of the pixels. Those properties attach to the image as a whole, at the higher level of description, and not to the pixels as seen through the microscope. It should be fairly clear that we will not be able to make sense of the image—for instance, to understand it as an image of the Mona Lisa—without attending to those holistic properties. In a similar vein, although an agent's intentional properties supervene on some underlying physical properties of the brain, they are not reducible to them. Supervenience can go along with irreducibility.

The Case for Realism about Intentional Agency

My argument for the claim that intentional agency is a real phenomenon rests on a form of scientific realism. At its centre is a methodological principle that is sometimes called the "naturalistic ontological attitude":

> **Naturalistic ontological attitude:** Our best guide to any questions about which entities, properties, or phenomena exist in any given domain is to be found in our best scientific theories of that domain (provided that, in scientific terms, we have no special reasons to doubt those theories).[31]

This principle tells us to take our best scientific theories at face value. If those theories imply that a particular phenomenon is real, and those theories are scientifically well corroborated, then there is no point in questioning whether the phenomenon is "really" real. For example, if we are interested in whether electromagnetic fields, gravitational forces, and the Higgs boson exist, we must consult our best theories of physics. If these theories say that those fields, forces, and particles exist, then we have every reason to treat them as real. To follow up by asking whether they are "really" real would be to ask one question too many. According to naturalistic realism, our ontology of the world must be guided by our best scientific theories. Of course, this means that we may sometimes have to revise our ontological hypotheses in light of new scientific findings, and this kind of contingency may not satisfy everyone's philosophical tastes. But if we want to determine which entities, properties, and phenomena are real, we simply have no better guide than science itself. As I have explained, our best theories in the human and behavioural sciences are committed to the view that there is such a phenomenon as intentional agency, so we have every reason to take that phenomenon at face value.[32]

Naturalistic realism stands opposed to instrumentalism. This is the view that whenever we postulate some entities, properties, or phenomena for explanatory purposes, those postulated items are best understood as instrumentally useful constructs, not necessarily as real—unless we can directly confirm their existence by observing them. Instrumentalism was prominently defended by the empiricists and logical positivists of the Vienna Circle, who thought that the main point of science was to accommodate our empirical observations.

An instrumentalist view about physics, for instance, would assert that the "unobservables" to which we frequently refer in order to make sense of our observations—from elementary particles to various fields and forces—are instrumentally useful constructs that we invoke in order to systematize those observations. It is useful to talk *as if* there are electromagnetic fields and gravitational forces, because this allows us to make sense of certain observable regularities, but we need not assume that those invisible fields and forces are themselves as real as tables, rocks, and armchairs (which are things that we can directly observe).[33]

Similarly, an instrumentalist view about psychology would assert the following: The observable behaviour of a human being or an animal is certainly real, and so is the physical activity in the brain; but the psychological states that we ascribe to an agent are merely instrumentally useful constructs that we employ to systematize the agent's observable behaviour. This kind of instrumentalism would amount to a form of "behaviourism." Only behaviour—something that we can directly observe—is truly real; the intentional properties we ascribe to agents are nothing more than instrumentally useful constructs; they are, at best, shorthand constructs for systematizing certain behavioural regularities.

Instrumentalist views in science, however, suffer from some well-known problems. They put far too much emphasis on the distinction between what is directly observable and what is not, and they adjudicate ontological questions—concerning what is or is not real—on the basis of that distinction. In practice, the boundary between observables and unobservables is vague and changeable. What counts as "observable" depends on the sophistication of our scientific instruments and on our measurement techniques. In chemistry and physics, tiny molecular structures that once seemed purely theoretical have become observable with the help of high-tech microscopes, for example. And our observations are themselves "theory laden"—that is, what we regard as empirical data will often come with certain theoretical presuppositions or auxiliary hypotheses. So, a philosophical view that makes the ontological status of an entity, property, or phenomenon conditional on whether it is observable relies far too much on a problematic distinction.

Furthermore, if we genuinely were to think that unobservable features such as electromagnetic fields and gravitational forces are nothing but theoretical constructs, we would face a major philosophical puzzle: Why do these constructs give us so much explanatory power? If electromagnetism and gravity truly exist, and if they play the roles that science says they do, then it is no miracle that postulating them allows us to explain our empirical observations. If, on the other hand, electromagnetism and gravity are nothing but theoretical constructs—figments of the observer's mind, not true features of the world—then the success of the relevant scientific theories seems a miracle. Arguably, a realist view about science—which acknowledges the reality of the unobservables—is "the only philosophy that doesn't make the success of science a miracle," as Hilary Putnam puts it.[34] This, in a nutshell, is the "no-miracles argument" for scientific realism. It seems, then, that realism is a more compelling philosophy of science than instrumentalism.

The same point can be made with regard to realism versus instrumentalism about intentional agency. If we genuinely were to think that the intentional agency we ascribe to people and complex animals is entirely in the eye of the beholder—nothing more than an instrumentally useful construct—then it would seem miraculous that we would be able to explain and predict human and animal behaviour so well on the basis of this ascription. Why should people and animals be so "usefully and voluminously predictable from the intentional stance" if their intentional agency is unreal? The no-miracles argument, which relies on an inference to the best explanation, supports the conclusion that intentional agency is a real phenomenon. In sum, intentional agency is no less real than electromagnetism, gravity, and other scientifically well-established phenomena. And this is not in tension with a naturalistic approach; to the contrary, it is supported by the naturalistic ontological attitude.

So, we are in a position to conclude that the challenge from radical materialism misfires. At least as far as the requirement of intentional agency is concerned, free will stands on scientifically acceptable foundations. Of course, we must also show that the other two requirements for free will can be met: alternative possibilities and causal control. Let us therefore move on.

4

In Defence of Alternative Possibilities

Our second requirement for free will is alternative possibilities: at least in relevant situations, a bearer of free will must have two or more possible courses of action. Each must be a genuine possibility for the agent. Without alternative possibilities, free will could not get off the ground. The idea is illustrated by Jorge Luis Borges's famous metaphor of the "Garden of Forking Paths." As we lead our lives, there are sometimes forks in the road ahead of us; we can choose between different futures, depending on how we act. My choice may be as simple as that between drinking a sip of water now and waiting a moment, and yet the future will be ever so slightly different, depending on how I choose. Or think of the more significant forks that we face: Should I become a teacher or try to pursue a better paid career in banking? Should I marry this person or not? Should I stay in my home country or try to emigrate?

To be sure, the difference we make is often rather small. It may be as small as that between drinking some water now and doing so a few minutes later. But still, the idea that different futures may be open to us is remarkable. After all, physics tells us that the world is governed by the underlying laws of nature. These determine the way the planets revolve around the sun, the way chemical reactions unfold, and how nature functions. Why should human choices fall outside those laws? Why should we face any forks in the road at all? Why should more than one future course of action ever be open to us?

As we have seen, the challenge from determinism is formidable. The thesis that the world is deterministic is often considered a hallmark of a scientific worldview. Determinism implies that, given the initial state of the universe, only one course of events will have been physically possible. If the world is deterministic, there could never be any alternative possibilities. The initial state of the world—say, at the time of the Big Bang—together with the laws of nature, would have been sufficient to determine all subsequent events, from the motion of the planets to all human behaviour. Your reading this book now would be nothing but an inevitable consequence of the physical past. You would never have had any genuine choices at all.

Now, you might say, the jury is still out on whether the world is deterministic. As already noted, there are strands in modern physics that depict the world as deterministic and others that depict it as indeterministic. We don't know yet on which side a grand unified theory of physics will fall, if it is ever discovered. It is unsettling to think, however, that the defensibility of free will might depend on how certain debates in physics play out. Should the idea of free will really hinge on how physicists resolve the tension between quantum mechanics and general relativity, for example? Currently these two fundamental theories are in conflict, and—under standard interpretations—they also fall on opposite sides in the determinism-indeterminism debate.[1] Furthermore, even if an indeterministic variant of physics were vindicated in the end, this would not automatically imply that we, human beings, have the kinds of alternative possibilities required for free will. It need not follow that physical indeterminism opens up degrees of freedom for us as agents rather than just introducing randomness into the world. Randomness alone would not secure free will.

My aim in this chapter is to argue that we, as human beings, do indeed have alternative possibilities. Against the challenge from determinism, I want to defend a claim that may seem surprising: Physical indeterminism is neither necessary nor even sufficient for alternative possibilities in agency. Alternative possibilities, I suggest, do require a form of indeterminism, but what they require is indeterminism *at the agential or psychological level, not at the physical one.* I will explain the distinction between these two notions, physical and agential indeterminism, and show that agential indeterminism is an emergent phenomenon, like agency itself. It is real, but located at a level distinct from physics. The world may or may not be deterministic at the physical level. What matters for free will is an open future at the level of agency. So, I must explain how this is compatible with a law-governed, even deterministic physical world.

What Does It Mean to Say That an Agent Has "Alternative Possibilities"?

I will begin by clarifying how we should understand "alternative possibilities."[2] To say that someone has alternative possibilities is to say that, when he or she chooses to perform a particular action, the actual action is not the only one open to this agent; he or she could also act otherwise. Our first task, therefore, is to spell out what "can do otherwise" means.

There are at least three kinds of interpretation on offer: the so-called conditional, dispositional, and modal ones. Each gives a different defining clause for the statement "the agent can do otherwise." The proposed defining clauses specify what condition must be met for an agent to have alternative possibilities under the given interpretation. The three proposals are as follows:

Conditional interpretation: If the agent were to try or choose to do otherwise, he or she would succeed.

Dispositional interpretation: The agent has the disposition to do otherwise when, in appropriate circumstances, he or she tries to do otherwise.

Modal interpretation: It is possible (in a sense to be spelt out further) for the agent to do otherwise.

The differences between these three proposals may seem subtle, so let's take a moment to compare them. Let's begin with the first one: the conditional interpretation. According to it, to determine whether an agent "can do otherwise" in a particular choice situation, we must figure out whether the following conditional is true: *if* the agent were to try or choose to do otherwise, *then* he or she would succeed. To establish whether this if-then statement is true, we must ask whether in the nearest possible worlds in which the antecedent clause is true (the "if" clause), the consequent clause is also true (the "then" clause). Schematically, a statement of the form "if *X* were the case, then *Y* would be the case" stands for the claim that in the nearest possible worlds in which *X* is true, *Y* is also true. By the "nearest possible worlds in which *X* is true" we mean, roughly, the most plausible and least far-fetched hypothetical such scenarios. In the present case, the antecedent clause is "the agent tries or chooses to do otherwise," and the consequent clause is "he or she succeeds." So, we must hypothetically conjure up the nearest possible worlds in which the agent tries or chooses to do otherwise and ask whether, in every such possible world, the agent succeeds. Whether the answer is yes or no will depend on what we take those nearest possible worlds to be and what we expect will happen in any one of them. Regardless of how we fill in those details, however, the criterion for alternative possibilities is a counterfactual one: it focuses on what would happen in a possible world distinct from the actual one.

Let's turn to the second proposal: the dispositional interpretation. Here, to determine whether an agent "can do otherwise" in a particular situation, we must first define what it means to have a *disposition* of the relevant kind—namely, the disposition to do otherwise in the appropriately qualified circumstances. We must then figure out whether the agent fits that requirement. We obtain different variants of this idea, depending on how exactly we define dispositions. The notion of a disposition, such as someone's disposition to react in certain ways in response to certain stimuli, is a topic of considerable complexity. Independently of the details, however, our criterion for alternative possibilities, on such an interpretation, is the presence of a suitable disposition in the agent. If the agent has the disposition, then he or she can be said to have alternative possibilities; if not, then not.

Finally, consider the third proposal: the modal interpretation. Here, in order to determine whether an agent "can do otherwise," we must ask what courses of action are *possible* for the agent: which future trajectories are accessible to him or her in the given situation. In other words, we must consider whether the agent faces a fork in the road. We end up with different variants of this interpretation, depending on how we spell out the notion of "possibility"—for instance, whether we mean physical, psychological, or some other kind of possibility. The modal interpretation is in line with the thought that I *can* do something only if it is *possible* for me to do it, in a relevant sense. Our criterion for alternative possibilities is then the presence of those possibilities.

Which interpretation of "can do otherwise" should we adopt? The first thing to note is this. Suppose, for the sake of argument, that we are interested only in reconciling alternative possibilities with determinism, not in adopting an interpretation that is independently plausible. If reconciling alternative possibilities with determinism is our only goal (perhaps even at the cost of gerrymandering our concepts), then a conditional or dispositional interpretation seems the one to go for. The desired reconciliation is much easier to achieve under either of these interpretations than under a modal one. To see this, consider the conditional interpretation. Even if I was completely predestined to do one thing rather than another, it can still be true that in a suitable hypothetical scenario in which I tried to do otherwise, I would succeed, and thus the conditional criterion for alternative possibilities can be met. The fact that determinism might have prevented that scenario from becoming actual makes no difference to the truth of the conditional statement. What matters for the conditional is what would have happened *if I had tried to do otherwise,* not whether I could have tried to do so. In short, alternative possibilities are compatible with determinism under the conditional interpretation. This point was recognized in the early 1900s by G. E. Moore, who defended free will by interpreting "can do otherwise" in this way.[3] He wrote,

> Our theory does not assert that any agent ever could have *chosen* any other action than the one he actually performed. It only asserts, that, in the case of all voluntary actions, he *could* have acted differently, *if* he had chosen: not that he could have made the choice.[4]

Of course, it remains a separate question whether this interpretation is independently plausible.

Similarly, consider the dispositional interpretation. Recall that, according to it, the criterion for having alternative possibilities is the presence of a suitable disposition in the agent—namely, the disposition to act otherwise in appropriate circumstances. Now, manifestly, *having* a disposition is not the same as *exercising* it. A precious vase may have the disposition of fragility: it is disposed to break under certain conditions; and yet, because we look after it carefully, it never actually breaks. The vase's fragility remains unexercised here. Analogously, even if I am somehow prevented from exercising the disposition to act otherwise, I can still have that disposition nonetheless. For this reason, determinism is compatible with alternative possibilities under the dispositional interpretation. This insight is central to an approach to free will that has become known as "the new dispositionalism."[5]

But now consider the modal interpretation. This makes it much harder to reconcile alternative possibilities with determinism. Suppose that having alternative possibilities requires that it be *possible* for an agent to do more than one thing in a given situation: multiple courses of action must be open. Then it is not clear how this can be squared with determinism. If the world is deterministic, then only one future sequence of events is possible given the initial state of the world, and so it seems that an agent could never face any forks in the road.

Although the conditional and dispositional interpretations look more congenial to a defence of alternative possibilities, I want to adopt the modal interpretation. The reason is simple: "can do otherwise" requires the genuine *possibility* of doing otherwise. The compatibility of determinism and alternative possibilities would come too cheap under a conditional or dispositional interpretation. We would be implausibly gerrymandering our notion of alternative possibilities. Recall the conditional interpretation. Suppose someone is totally incapable of *trying* to act differently than he or she actually does, perhaps due to some psychological compulsion. Then it seems implausible to say that he or she genuinely has alternative possibilities. Yet, under the conditional interpretation, the person might still count as having "alternative possibilities": it can be true that in a hypothetical scenario in which the person tries to do something else, he or she would succeed. The fact that such a scenario could never materialize does not undermine the truth of the conditional statement. This suggests that having "alternative possibilities" in the conditional sense does not match what we ordinarily mean by "being able to do otherwise."

A similar point can be made about the dispositional interpretation. A pianist might be disposed to play a particular Mozart sonata flawlessly in normal circumstances, but might be psychologically bound to freeze under the special pressure of an audition. In this case, our pianist could not perform any better in the audition. He lacks alternative possibilities *in the given situation,* despite having the right general disposition.[6] The disposition alone seems insufficient for the presence of alternative possibilities. As Ann Whittle notes, "The dispositional analyses of abilities . . . latch on to [a] global sense of ability. But such global abilities to do otherwise do not capture the kind of freedom that is necessary for moral responsibility."[7]

In sum, both the conditional and the dispositional interpretation offer only a watered-down notion of alternative possibilities: a notion that does not capture what we normally mean by saying that someone "could have done otherwise."[8] I could have chosen tea rather than coffee. A violent criminal could have refrained from committing his crime. I could have pursued a different career. I could have helped the stranger in need. And so on. In all those cases, alternative possibilities must literally mean alternative *possibilities.* An agent can do otherwise only if it is *possible* for the agent to do otherwise. This is precisely what is implied by the modal interpretation. Susan Hurley makes the same point:

> The ability to do otherwise entails the *outright possibility* of acting otherwise: it entails that there is a causal possibility of acting otherwise, holding all else constant. A counterfactually conditioned disposition to act otherwise is not the same thing as an outright possibility of acting otherwise. . . . That the former is compatible with determinism does not entail that the latter is.[9]

Now that we have seen how to interpret "alternative possibilities," we can return to the challenge that determinism poses for free will. If free will requires that more than one course of action be *possible* for the agent, is there any hope of reconciling free will with determinism?

The Challenge Revisited

It is useful to restate the argument for the incompatibility of free will and determinism:

> **Premise 1:** Free will requires that, at least in relevant situations, more than one course of action be possible for the agent.
> **Premise 2:** Determinism implies that, in any situation, only one course of action is ever possible for the agent.
> **Conclusion:** Free will and determinism are incompatible.

From a purely logical perspective, this argument is clearly valid. If we accept the premises, we must also accept the conclusion. Furthermore, the argument looks sound, since the premises seem hard to deny. The first premise is a restatement of the requirement of alternative possibilities, interpreted in the modal terms for which I have argued. And the second premise seems a straightforward statement of what determinism implies. If the laws of nature are deterministic, the initial state of the world leaves no room for more than one course of events, and so an agent could never do anything other than what he or she actually does. The upshot is that determinism rules out free will.

On closer inspection, however, this line of reasoning is too quick. Recall the notion of determinism that features in Newtonian physics and in Pierre Laplace's thought experiment of the demon who can predict the future from the past. This notion is a physical one. It refers to the fundamental laws of physics. We can define it as follows:

> **Physical determinism:** Given the complete physical state of the world at any point in time, only one future sequence of events is physically possible. "Physically possible," in turn, means "compatible with the fundamental physical laws."

This thesis captures the sense in which Enlightenment science in Isaac Newton's tradition depicts nature as deterministic. And it captures the idea behind Albert Einstein's famous comment, "God does not play dice." A satisfactory microphysical theory, Einstein thought, should depict nature as deterministic.[10]

Crucially, physical determinism is a thesis about *physical possibility.* This means that the wording of premise 2 in the argument above is not quite accurate. As currently worded, premise 2 construes determinism as a thesis about *possibility for the agent*—namely, as the following:

> **Agential determinism:** In any situation, only one course of action is possible for the agent.

This is not the same as physical determinism, and so, if the point of premise 2 is to capture the kind of determinism asserted by classical physics, then the following wording would be more accurate:

> **Premise 2 (reworded):** Determinism implies that, given the complete physical state of the world at any point in time, only one future sequence of events is physically possible.

But what happens if we replace the original premise 2 with this reworded one? Would the incompatibility of free will and determinism still follow? Here is what the amended argument would look like:

> **Premise 1:** Free will requires that, at least in relevant situations, more than one course of action be *possible for the agent.*

Premise 2 (reworded): Determinism implies that, given the complete physical state of the world at any point in time, only one future sequence of events is *physically possible.*

Conclusion: Free will and determinism are incompatible.

From a logical perspective, this argument is no longer valid. There are now two senses of possibility involved: agential and physical. According to premise 1, free will requires certain possibilities in one sense: an agential one. According to the reworded premise 2, determinism rules out certain possibilities in another sense: a physical one. Given these different senses of possibility, no conflict between free will and determinism can be directly inferred from this, and the original conclusion no longer follows.[11]

Now, you might say, Wait a minute; although the reformulated argument is not valid in its present form, its validity can easily be restored by adding a missing premise. Suppose we could demonstrate the following:

> *If,* given the complete physical state of the world at any point in time, only one future sequence of events is physically possible, *then,* in any situation, only one course of action is ever possible for an agent.

Call this the "linking thesis." At first sight, it sounds plausible. *If* physical determinism holds, *then* it may seem that agential determinism must hold as well. If we accept this linking thesis, then the incompatibility of free will and determinism can once again be derived. Premise 1 and the reworded premise 2, together with the linking thesis, entail the incompatibilist conclusion. If physical determinism implies agential determinism, and agential determinism rules out free will, then physical determinism also rules out free will. So, the challenge for free will stands.

What can be said in response? In what follows I will show that, initial appearances notwithstanding, the "linking thesis" is false: physical determinism does not imply agential determinism. In fact, the original argument for the incompatibility of free will and determinism involves a kind of "category mistake": a mixing of two different levels of description that do not fit together.[12]

Why Agential Indeterminism Is Compatible with Physical Determinism

I have argued that free will requires a form of indeterminism—namely:

> **Agential indeterminism:** In relevant situations, more than one course of action is possible for the agent.

I want to show that this kind of indeterminism is not threatened by physical determinism; physical determinism does not imply agential determinism.[13]

The first thing to observe is that physical and agential determinism are defined at two different levels of description. One is a thesis at the physical level, the other a thesis at the psychological one: the level at which we speak about agents and their intentional actions. Let me briefly revisit the idea of "levels" and then tease agential and physical determinism apart.

As I have pointed out, different domains of enquiry describe the world differently. When we investigate the basic laws of nature, we use fundamental physical descriptions. We employ concepts and categories such as particles, fields, and forces, and we use equations to express how physical systems evolve over time. When instead we study chemistry or biology, we use different concepts and categories, ranging from molecules and chemical reactions to cells and organisms. These give rise to a higher level of description, one that abstracts away from fundamental physics. Fundamental physical descriptions are absent from chemistry and biology, and this is for a good reason: chemical and biological explanations would not benefit from extraneous physical details. Such details would add clutter or create noise rather than capture chemically or biologically relevant patterns or information.

Now, when we turn to psychological and social phenomena, we abstract even further away from fundamental physics and move to yet another level of description, employing concepts such as intention and action, belief and desire, goal and purpose. I will call this the "psychological level." The psychological level, roughly speaking, is the level at which our best theories of psychology describe and explain human cognition and behaviour. Just as it would be hopeless to explain a word processor at the hardware level at which millions of electrons flow through microchips, so it would be hopeless to explain human behaviour in fundamental physical terms. Recall that the hallmark of any higher-level phenomenon is that it can be adequately captured only by higher-level descriptions. Intentional agency, as we have seen, is quintessentially a higher-level phenomenon. Any talk of agents and their intentional states and actions is not reducible to lower-level talk of physical processes in the brain and body. (Importantly, what I have called the "psychological level" must not be confused with a person's subjective, first-personal perspective. What people subjectively experience need not coincide with how our best theories of psychology describe human cognition and behaviour; it is the latter that the term "psychological level" refers to.)

What all this implies is this: When we are asking what an agent can or cannot do, the right level of description is not the fundamental physical level but the psychological one. The question of whether *agential* determinism holds is indeed relevant to what an agent can or cannot do, because agential determinism pertains to the psychological level. But the question of whether *physical* determinism holds is not relevant. Or even if it were relevant, its relevance would be indirect, and it would require further argument. Physical determinism is a thesis about a level well below that of psychology; we cannot speak about agents and their intentional actions at the physical level. It is only *agential* determinism that would pose a threat to free will, if it were vindicated.

So far we have seen that physical and agential determinism are distinct theses. They refer to two very different levels of description. I now want to show that they are not just distinct, but logically independent: neither thesis implies the other. For our purposes, it is particularly important that there can be physical determinism without agential determinism. I will first sketch the argument informally. I will then introduce a little formal model to establish my point more precisely.

The argument as to why physical and agential determinism may come apart begins with a structural point already discussed: Physical and agential phenomena stand in a relation of *supervenience with multiple realizability.* This means that the same psychological properties of an agent can be instantiated by different configurations of physical properties of the underlying organism. Different microlevel patterns of neuronal activity in a person's brain may instantiate, for example, the same intentional property of desiring to drink some water. Similarly, the belief that Washington, DC, is the capital of the United States can be neuronally encoded in a gazillion different ways. In that sense, psychological properties are more abstract and coarse-grained than neural properties, let alone microphysical ones. To describe a person's properties in psychological terms, we need not describe all lower-level properties. A person's beliefs, desires, intentions, memories, and other psychological properties may be compatible with different subvenient physical details. Specifying those details would distract us from capturing the psychological properties of interest.

This point has a crucial implication: Even the totality of all psychological facts at a given point in time may be consistent with a variety of different physical states of the world. A complete specification of all agents' psychological states and their macroscopic environment does not entail a complete specification of all the underlying microphysical details. Different physical configurations—most likely an inordinate number—can instantiate the same state of the world *as specified through the lens of psychology.* Those different physical configurations are then indistinguishable *in psychological terms.* From a higher-level perspective, they are completely equivalent. We may express this equivalence by saying that, at the psychological level, there is no fact of the matter as to which precise physical state obtains.[14]

This point remains true even when psychological-level descriptions refer to the physical environment in which human action takes place. Of course, psychological-level descriptions refer not only to people's mental states and cognitive processes. They refer also to what people do with their bodies, and these, in turn, interact with the physical environment. Crucially, however, when we refer to the physical objects and external resources involved in human action, and we do so through the lens of the behavioural and social sciences—for instance, when we refer to the means of production in economics or the material environment in sociology or psychology—we use macroscopic, coarse-grained descriptions, not microscopic, fine-grained ones like in fundamental physics.

The coarse-grained nature of the psychological level opens up the possibility that the state of the world as specified at that level may be consistent with more than one sequence of events, even if there is physical determinism. In particular, a psychological-level state is consistent with every sequence of events that is supported by one of its possible physical realizations. Remember that the higher-level state, at any time, does not determine which underlying lower-level state obtains, and so we cannot treat any one of these lower-level states as the "true" one. To do so would be to abandon the higher-level perspective and to step down to the lower level.

As long as some of the possible higher-level sequences of events correspond to different courses of action, it follows that more than one course of action is *possible for the agent.* In short, the totality of facts at the psychological level up to a given time may leave more than one future course of action open for an agent. And so, there may be agential indeterminism, even in the presence of physical determinism. This completes the informal version of the argument. But since this may have been a little quick, I will now make the argument more precise, with the help of a simple formal model.[15] Readers who are satisfied with the informal version of the argument should feel free to skim or even skip the rest of this section.

Think of the world—or the part of the world we are concerned with—as a dynamic system. Familiar examples of such systems are the solar system, the climate system, or the financial system. At each point in time, a dynamic system is in a particular state, and that state may evolve over time on the basis of certain laws governing the system. In the case of some systems, such as the solar system, those laws may be deterministic;

in the case of others, such as the weather or climate, they may be probabilistic (in this case, we call the system "stochastic"). For the purposes of our model, time is linearly ordered. Call the set of all possible momentary states in which the system could be its "state space." A *history* of the system is a trajectory of the system through its state space, as time passes. The laws of the system can then be understood as the constraints specifying which histories are possible—we shall say *nomologically possible*—and which are not. In our model, the laws can thus be captured by defining the set of nomologically possible histories of the system.[16]

For example, the nomologically possible histories could be as shown in Figure 1. In this simple illustration, there are six time periods, from time 1 to time 6. Time 1 corresponds to the bottom row of the diagram, time 6 corresponds to the top row. Dots represent states of the system, and lines represent histories. Each history thus begins in the bottom row and continues upwards. We can think of the state at time 1 as the system's initial state, which is followed by a sequence of states at subsequent times. In Figure 1, all the nomologically possible histories are deterministic: the initial segment of each history fully determines the rest of that history. Indeed, once we fix the state at time 1, there is only one possible way the history can unfold. Different states at time 1 correspond to different initial conditions.

Technically, a history is called "deterministic" if, at every time, its initial segment up to that time admits only one possible continuation under the laws governing the system. Graphically, the defining feature of a deterministic history is the lack of any branching. In an *indeterministic* history, by contrast, there would be branching: the initial segment of such a history would not always determine the rest of the history; two or more distinct continuations may be possible. If our physical universe behaves structurally like one of the histories in Figure 1, then the thesis of physical determinism is true. Of course, whether our universe is like this is an open question. So far, so good.

Now, assuming we treat the states of our model system as physical states, then each state can be interpreted as a full specification of all physical particles, fields, and forces at a particular point in time. Accordingly,

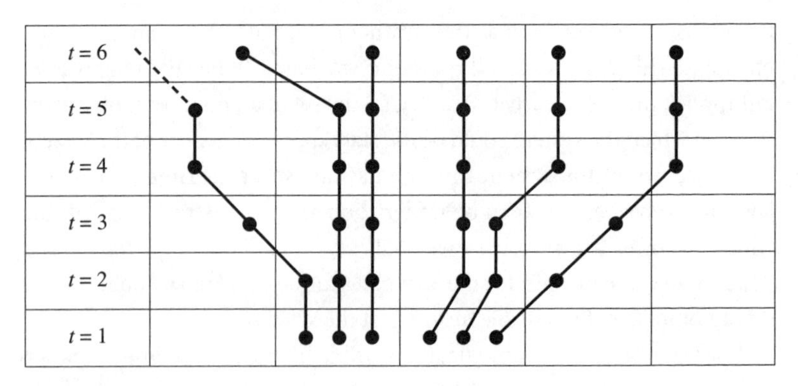

Figure 1. Deterministic lower-level histories

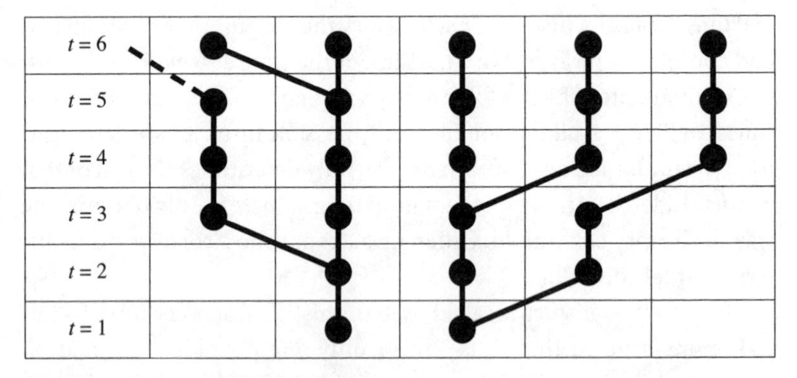

Figure 2. Indeterministic higher-level histories

histories can be interpreted as physical histories. Suppose, however, that we are interested not in how the system behaves at the fundamental physical level but in certain higher-level phenomena. Specifically, suppose that each physical state gives rise to a resulting higher-level state—namely, a psychological-level state. This specifies all the psychological properties of the relevant agents and their macroscopic environment. The higher-level state supervenes on the underlying physical state. Suppose further, as discussed, that higher-level states are multiply realizable: different physical configurations can instantiate the same higher-level state. For concreteness, suppose that whenever two or more distinct physical states fall inside the same cell within the rectangular grid in Figure 1, they realize the same higher-level state. For instance, the six possible physical states at time 1 correspond to only two possible psychological-level states: one corresponding to the occupied cell on the left-hand side, the other corresponding to the occupied cell on the right. The physical states within each of these cells are indistinguishable as far as the relevant macroscopic properties go, especially with respect to all psychological properties.

Let's see what our system looks like when we redescribe it in these higher-level terms, focusing no longer on physical states and histories but on psychological ones. Figure 2 shows the states and histories under this redescription. Thick dots represent higher-level states, and thick lines represent higher-level histories. The most striking feature of Figure 2 is that the histories displayed here are indeterministic. It is no longer true that the initial segment of any history always determines the rest of that history. Instead, there is some branching. Irrespective of the initial state at time 1, there are always several possible ways in which the history can unfold. Higher-level indeterminism arises as an emergent by-product of lower-level determinism. Note that we are not cheating by changing the definitions of determinism and indeterminism. As in the lower-level case, determinism of a given history of the system still means that this history's initial segment up to any point in time admits only one possible continuation under the relevant laws; and indeterminism means the opposite. We are now simply applying those definitions to higher-level histories, rather than lower-level ones.

What these considerations show is that lower-level determinism is compatible with higher-level indeterminism, and so determinism at the physical level can coexist with indeterminism at the level of agency. Of course, taken in isolation, the present model offers only a proof of

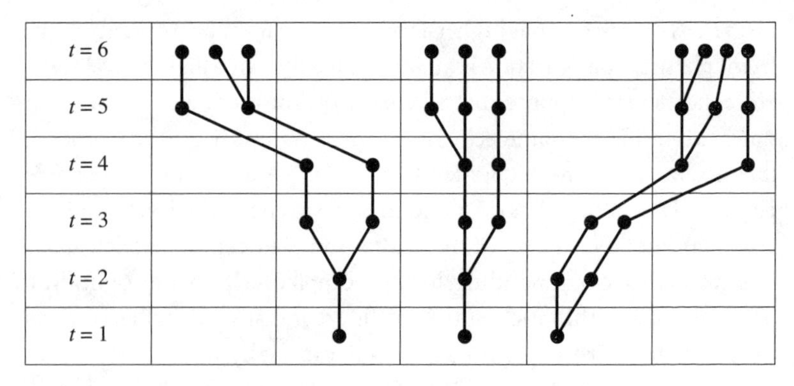

Figure 3. Indeterministic lower-level histories

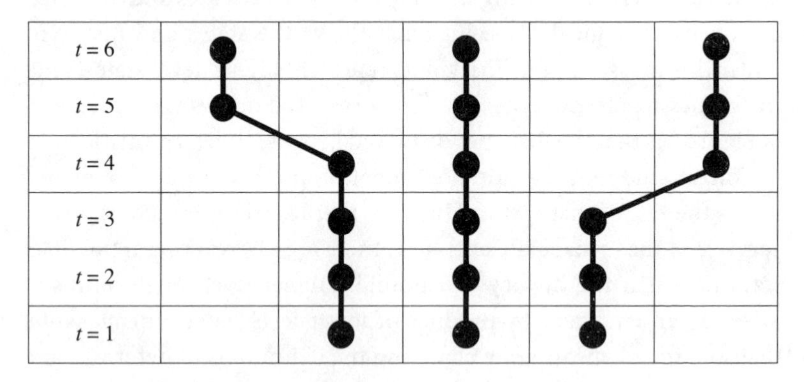

Figure 4. Deterministic higher-level histories

concept. We do not know whether physics is deterministic; nor have I argued yet that psychology is indeterministic. We can conclude, however, that what I have called the "linking thesis" is false: physical determinism does not imply agential determinism.

It is worth noting that the reverse scenario is coherent as well: there can be lower-level *indeterminism* together with higher-level determinism. If the lower-level histories were as shown in Figure 3, for example, then the resulting higher-level histories would be as shown in

Figure 4. Figure 4 is derived from Figure 3 by redescribing states and histories in the same way as explained before. Again, thick dots represent higher-level states, and thin dots represent the underlying lower-level states. As before, all the lower-level states in the same cell give rise to the same higher-level state.

In this example, there is branching in lower-level histories but no branching in higher-level ones. This might correspond to a scenario in which lower-level histories behave according to indeterministic laws of quantum mechanics (Copenhagen style), while higher-level histories behave according to Newton's deterministic laws. Here the effect of supervenience and multiple realizability is the opposite of the earlier one. In short, just as lower-level determinism can coexist with higher-level indeterminism, so lower-level indeterminism can coexist with higher-level determinism. As philosophers of physics recognize, it can easily happen that a system's "micro- and macro-dynamics do not mesh," to use Jeremy Butterfield's words.[17]

This completes my argument for the claim that, logically speaking, physical and agential determinism are independent from one another. In consequence—and this is a really important point—physical indeterminism is neither necessary nor even sufficient for alternative possibilities in agency. We must ask, then, whether we have good reason to accept the thesis of agential indeterminism. The key issue, I will suggest, is whether our best theories of agency support such agential indeterminism.

The Case for Realism about Alternative Possibilities

It should be clear by now that determinism at the fundamental physical level does not challenge free will. It leaves the possibility of agential indeterminism entirely intact, and only the latter—not physical indeterminism—is required for free will. By contrast, determinism at the level of psychology would pose a very significant challenge. If our best theories of human agency were to show that people are deterministic systems, it would be "game over" for the attempt to defend free will in the libertarian sense I am trying to defend. However, my claim is the following:

> **Indeterminism in psychology:** Our best theories of human agency, in areas ranging from cognitive psychology to behavioural economics, presuppose that people face choices between more than one course of action. For these theories, an important goal is to explain how people make such choices, not to explain the phenomenon of choice making away.

That psychology and the behavioural and social sciences do not depict human beings as deterministic is a crucial premise for my defence of free will.[18] I am betting on the truth of this claim not just for now but also in relation to future scientific theories in psychology and the social sciences. In particular, I hypothesize that even our best future theories of human agency will retain the presupposition of agential indeterminism. If scientists discovered deterministic laws of human psychology, akin to Newton's laws governing the motion of the planets, and if these laws allowed us to dispense with our understanding of people as choice-making agents, I would have to concede defeat. I announced at the outset that my defence of free will would rely on certain empirical premises. The thesis that psychology supports agential indeterminism is one such premise. If this premise is true, however, we are warranted in taking a realist view about agential indeterminism—that is, we are warranted in considering it a real phenomenon; or so I will argue.

Now, I should acknowledge that, in both public and scientific discourse, we keep hearing claims to the effect that certain forms of human behaviour are explained by people's genes, their social and economic backgrounds, their upbringing and education, their peers, and other factors. But those research findings are rarely able to attribute more than a certain *part* of the observed behavioural variation across people to those purported explanatory variables. The variables in question are typically shown, at most, to affect the *probabilities* of certain behaviours, without fully determining them, and there is little reason to think that we will ever arrive at a truly deterministic theory of psychology.[19]

As things stand right now, we would not even know where to begin if we tried to explain human behaviour without assuming that people face genuine choices, with several options in front of them. To be sure, psychologists and social scientists would like to explain and predict human behaviour better than we are currently able to do. We would like to develop a better understanding of the determinants of consumer choices, of voting decisions, and of people's dispositions to cooperate in strategic interactions, for example. In most of these cases, however, our explanations proceed by granting that there are several options among which an agent could in principle choose, and then offering an account of which ones among these options the agent is likely to choose and why. Similarly, our attempts to predict an agent's behaviour typically begin with the assumption that the agent has several options, and we then try to predict his or her choice among these possibilities by identifying which options stand out as rational for the agent, psychologically salient, or favoured by the agent's choice heuristics. Regardless of the details, the actual choice is usually assumed to be drawn from a menu of possibilities.

I have already argued for the indispensability of agency ascriptions for explaining and predicting human behaviour. I now want to suggest, further, that the presupposition of agential indeterminism is central to the logic of this enterprise. An intentional explanation usually proceeds as follows: In a given choice situation, an agent is faced with a number of in-principle possible options. This range of possibilities may then be reduced by certain feasibility constraints or other considerations of psychological salience. Finally, some cognitive criterion—whether a criterion of rationality or something less rationalistic—is applied in order to identify the option(s) that the agent will choose. The distinction between "possible options" and "chosen options" is essential for the logic of any such explanation. The key assumption is that, while the agent will of course choose just one option in the end, the others would have been possible too, in some sense of "possibility."[20] If we denied that there were other possible options distinct from the chosen option, we would end up trivializing intentional explanations.

To reinforce this point, consider a very simple theory of choice. This theory, which I am bringing up only for illustrative purposes, asserts that

- each agent has a fixed preference ordering over all the options that might come up; and
- in any situation, he or she chooses the most preferred option among the options that are possible in that situation.

This theory is a very crude version of the "preference maximization theory," which used to be influential in economics before the arrival of more psychologically informed theories. Most psychologists will consider the crude preference maximization theory perfectly intelligible, but they will think—quite rightly—that it is falsified by what we know about human psychology. It is falsified, among other things, because we cannot generally explain all choices by attributing to each agent a single, completely fixed preference ordering that is independent of the context of choice.[21] However—and this is the point I want to make—if we seriously assumed that *only one* option is possible in each choice situation due to agential determinism, then this would render the theory *vacuously true.* Irrespective of the agent's preferences, if there is only one possible option for the agent in each situation, then this will automatically be the most preferred option in that situation; there are no others. Our crude theory would then come out as *true, albeit trivially so.* Instead, we actually think that the theory is *false but nontrivially so.* Its falsity has been established via empirical investigation. It took some psychological experiments, including the famous framing experiments conducted by Daniel Kahneman and Amos Tversky, to falsify a theory like the present one.[22]

What's more, under the assumption of agential determinism, our illustrative theory would have to be judged not just trivially true, but indistinguishable from the theory that says that the agent will always choose the *least* (as opposed to *most*) preferred option, or from the theory that says that the agent will always randomly pick one option or use some other heuristic. If only one option were genuinely available in each situation, then all these different theories would be *de facto* equivalent. The very fact that scholars treat them as distinct theories and compare them empirically shows that these scholars cannot be assuming that people's sets of possibilities are always singleton, as agential determinism would entail.

It seems, then, that the assumption of agential indeterminism is at the heart of intentional explanation: there must be a sense of *possibility* in which an agent faces choices among several possible options; the goal is to identify the agent's chosen options among the possible ones. It is even unclear to what extent we would be warranted in ascribing agency to an entity in the first place unless we took that entity to have alternative possibilities. In a similar spirit, Helen Steward argues that the very notion of agency presupposes alternative possibilities.[23] As I would put it, without alternative possibilities, there could be no agency in anything more than a trivial sense.

Consistent with this, the assumption of agential indeterminism seems to be shared by practically all theories of intentional agency, ranging from the crudest versions of rational choice theory to the most sophisticated theories in behavioural economics and cognitive psychology. For example, a key notion in decision and game theory is that of *an agent's set of possible actions or strategies.* The explanation of many social phenomena relies on the assumption that agents' action or strategy sets contain more than one option. Sometimes the addition or removal of options can make a difference to what an agent is predicted to do even if the added or removed options are never actually chosen. Unless we accept that there is a sense—at least a technical one—in which those options are genuinely available, it is hard to account for those effects.[24]

Could we still adopt an instrumentalist view and suggest that the postulated option sets are nothing but explanatorily useful constructs, which do not capture any real possibilities? This interpretation would, once again, parallel the instrumentalist view in the philosophy of science, according to which unobservables such as electrons and magnetic fields are nothing but instrumentally useful constructs, which should not be viewed as real. In the present case, the unobservables would be the option sets we attribute to an agent, which usually include some options that the agent will not choose. We would be postulating these sets of so-called possible options merely to make sense of certain regularities in the agent's observable behaviour, but we would not treat them as genuine possibilities. They would be instrumentally useful constructs of our explanations, in the same way in which certain unobservable elementary particles and fields are instrumentally useful constructs of fundamental physics.

As before, however, such an instrumentalist view would go against the naturalistic ontological attitude that I have assumed throughout this book. If we hold on to that attitude, then we must take the ontological commitments of our best scientific theories in any given domain at face value, provided we have no special reasons to doubt those theories. This is the same attitude that physicists take when they view the Higgs boson and other elementary particles as real, given that our best theories of fundamental physics are committed to their existence.

You might object that the analogy between the theory of agency and physics is not plausible. After all, the Higgs boson and other elementary particles—unlike a human decision maker's unchosen options—have been empirically observed, using the Large Hadron Collider experiment at CERN (the Conseil Européen pour la Recherche Nucléaire). My response, however, is that the analogy holds up. What the experimental work at CERN has given us is not a *direct observation* of the particles in question but merely a large data set which can be *best explained* by postulating those particles. An inference to the best explanation is still required to support a realist attitude towards those entities. Similarly, if we accept that our best theories of agency are committed to nonsingleton sets of options (and we grant that these theories are not fundamentally in doubt), then we have every reason to treat agential indeterminism as a real phenomenon.[25]

Now, there is one further objection that we need to consider. Even if we reject the kind of instrumentalist view I have just discussed, might we still argue that the postulated options are not real possibilities but merely *imagined possibilities* in the minds of the agents reasoning about them? This view would shift the possibilities in question from the world itself to the minds of the agents we are trying to explain. In particular, the "possible" options would then be those that an agent *subjectively believes* to be possible, even though in reality there is only one genuinely possible option in each situation—namely, the one that ends up being chosen.

Perhaps we could render such an interpretation coherent, but it would be more complicated and revisionary than the more literal interpretation according to which the possible options are the possibilities the agent is genuinely faced with. The revisionary interpretation would have to attribute to each agent a mental state in which, although there are no real choices, the agent systematically conjures up a set of imaginary possibilities of which only one is a genuine possibility. This would amount to a kind of "error theory" concerning the nature of human deliberation: agents would be systematically mistaken in thinking that they are faced with possible choices when in fact they never have more than one option.

If physical determinism were to undermine the idea that human beings have alternative possibilities, then we might have some reason to go for this revisionary interpretation. After all, the revisionary interpretation might then offer a way of reconciling the *appearance* of choice with its lack of reality, against the background of determinism. However, as we have seen, even if the world is deterministic at the physical level, this does not rule out agential indeterminism. So we can unproblematically adopt the simpler interpretation, according to which agential indeterminism is a real phenomenon.

Why My Defence of Alternative Possibilities Is Compatible with Ordinary Language

My focus has been on what our best scientific theories of agency say about alternative possibilities, rather than on our commonsense understanding of this idea. However, I would like to take a moment to explain why my analysis is consistent with the way we ordinarily speak about what a person can or cannot do.[26] To do so, I would like to relate my account of alternative possibilities to a classic account of the ordinary meaning of the word "can." That account was developed in the 1970s by the linguist Angelika Kratzer and has inspired much work in linguistics and related fields.[27]

According to Kratzer, the verb "can" in a sentence such as "A can do B" always comes with an additional, often implicit qualification of the form "in view of *X*." This qualification, if not specified explicitly, is determined by the context in which the sentence is uttered. For example, when we talk about what is humanly possible, we might say "We can walk, run, and jump, but not fly," thereby referring to what we can do in view of the constraints of human physiology. When we talk about how to get from London to Paris, we might say "We can fly, take the Eurostar train, or travel by overnight bus," thereby referring to what we can do in view of the available means of transport. We might add, however, "We cannot fly or take the Eurostar, because those options are too expensive," this time referring to what we can do in view of our financial constraints.

As these examples show, different contexts in which we use the word "can" give rise to different meanings. They do so by making different constraints salient, relative to which the word "can" is to be understood. Just as the earlier examples illustrate certain physical, social, or economic meanings of "can," so other contexts can give rise to logical, normative, or epistemic (knowledge-related) meanings. Examples are "I cannot square the circle" (logical), "You cannot step onto the lawn—it's forbidden" (normative), and "For all we know, the burglar can be brown-haired or red-haired" (epistemic).

In general, Kratzer suggests that the sentence "*A* can do *B* in view of *X*" is true if and only if *A*'s doing *B* is consistent with the constraints picked out by *X*.[28] Weaker constraints thus render more things possible, stronger constraints fewer things, and it may depend on the context of utterance whether it is true to say that someone can do something. The relevant constraints will settle that question one way or another.

Returning to the issue of alternative possibilities, when we talk about what an agent can and cannot do, whether in ordinary discourse or in science, the constraints relative to which this assertion is to be understood are given by our appropriate conception of agency, not by fundamental physics. This is certainly so in ordinary discourse. Consider the sentence

(S) "Brutus could have chosen not to murder Caesar."

The correct interpretation is not

> (S-physical) "Brutus could have chosen not to murder Caesar in view of the full microphysical history of the world up to the act in question, including his own neurophysiological history."

Rather, it is

> (S-psychological) "Brutus could have chosen not to murder Caesar in view of his capacities as an agent."

The physical interpretation would be incorrect because it would fail to pick out the right qualification of the word "can." We are clearly taking an intentional stance towards Brutus, and we should therefore be focusing on Brutus's capacities as an agent, not on the microphysical backstory. The physical interpretation of our sentence would involve an incongruous mixing of physical and agential descriptions. The psychological interpretation avoids this mistake.

The same point applies also to more scientific contexts—say, in psychology or decision theory. Here, too, the psychological interpretation of our sentence is the correct one. The difference between ordinary and scientific contexts lies in the level of sophistication with which we determine what an agent's capacities are—that is, whether we employ folk psychology or our most advanced versions of psychological decision theory for identifying those capacities. But despite this difference there is no reason to think that in scientific contexts the correct interpretation of our sentence suddenly shifts from the psychological to the physical one. These observations reinforce my insistence on interpreting alternative possibilities as a psychological rather than physical phenomenon.

Kratzer herself makes a similar observation, which is worth quoting at length. For present purposes, I will amend a few details, as indicated in square brackets:

> Consider the following case of a misunderstanding. Last year I attended a lecture . . . given by a man called "Professor Schielrecht"[, . . .]

> a third-generation offspring of the Vienna Circle, [whose] main concern in philosophy is to show that most of what most people say most of the time does not make sense.
>
> Suppose a judge asks himself whether a murderer could have acted otherwise than he eventually did. Professor Schielrecht said that the judge asks himself a question which does not make sense. Why not? Professor Schielrecht's answer was: Given the whole situation of the crime [my substitution: "given the full microphysical history leading up to the crime"], which includes of course all the dispositions of the murderer [my substitution: "all relevant neurophysiological processes leading up to the murderer's action"], this man could not have acted otherwise than as he did. . . .
>
> [But t]he answer to the question of the judge is *not* trivial. The judge asked himself: Could this murderer have acted otherwise than he eventually did? . . . The judge did not make explicit the first [qualification] required by the word "could" which he used in his question. Professor Schielrecht provided such [a qualification] by the phrase "given the whole situation" [my substitution: "the full physical backstory"], but [this] does not match what the judge meant. He misunderstood the judge: what the judge probably meant was: Given such and such aspects of the situation [my substitution: "given our best psychological description of the situation"], could the murderer have acted otherwise than he eventually did?[29]

In line with Kratzer's interpretation, I suggest that it is the latter sort of question that matters for free will, not the question Professor Schielrecht incorrectly attributed to the judge. What matters is whether, given our best psychological description of the situation, the murderer could have acted otherwise, not whether the murderer, reconstrued as a purely physical organism, could have ended up on a different physical trajectory given the complete microphysical history of the world. Indeed, we usually consider a positive answer to the judge's question under Kratzer's interpretation an appropriate condition for moral and legal responsibility, while we deny the relevance of Professor Schielrecht's interpretation.

Consider what kind of evidence a judge would accept as exculpatory evidence in a criminal trial. Suppose Jones is accused of killing Smith. If Jones turns out to have a lesion in his brain that significantly affects his agential capacities—for instance, by systematically failing to inhibit his aggression—then this is relevant to the question of whether Jones should be held criminally liable. In particular, if Jones's medical doctor testifies that Jones's action was the inevitable result of a neurological disorder or psychological compulsion—say, because of certain agency-constraining effects of the lesion—then the judge would conclude that Jones's act was due to an illness that is beyond his control. He could not have acted otherwise, and should therefore not be convicted for murder but instead sent to the hospital for medical treatment.

By contrast, suppose that Jones's brain and body are completely normal, and there is no health condition affecting his agential capacities. Now imagine Jones's defence lawyer brings in a theoretical physicist as an expert witness. The physicist testifies that, due to microphysical determinism, Jones's behaviour was predetermined at some fundamental physical level. I suspect that there is no way the judge would let Jones off the hook. Most likely, the judge would find the physicist's testimony completely irrelevant to the case. Why? Because the testimony targets the wrong level of description. What the judge is interested in is Jones's conduct as an agent, not the microphysical trajectory of the underlying bodily system. My account of alternative possibilities vindicates the judge's perspective. The kinds of alternative possibilities that matter for free will are agential-level possibilities, not physical-level ones.

Forks in the Road, and Agency

I have defended free will against the challenge from determinism. Human beings can have alternative possibilities as an emergent higher-level phenomenon, even in a deterministic physical world. Free will requires *agential* indeterminism, not *physical* indeterminism. Physical indeterminism is neither necessary nor sufficient for agential indeterminism, and so the question of whether science will discover determinism at the physical level is much less relevant to the free-will debate than often assumed. By contrast, determinism in psychology would pose a massive challenge for free will. Fortunately, however, our current best theories of agency support agential indeterminism, and I have suggested that we have good reasons to expect that future theories in psychology will retain this feature.

Now, if we accept the view that human beings face genuine forks in the road, we might worry that there is another challenge for free will looming on the horizon. If an agent's action is left genuinely undetermined by his or her prior psychological state, including the agent's beliefs and desires, then we might wonder how this action could qualify as something for which the agent is truly responsible. How could the action be genuinely attributed to the agent, and not to some random or otherwise indeterminate process?[30]

In particular, if we grant that

1. more than one course of action is possible for the agent, even given the agent's full psychological state,

then the worry is that

2. the agent's actual action is an agentially undetermined fluke event for which the agent is not responsible.

While claim 1 is the desired implication of agential indeterminism, claim 2 seems to be the flip side. If several distinct courses of action are equally consistent with the agent's full psychological state at the given time, then it is hard to see how the agent's actual action could be any more attributable to the agent than any of the other possible alternatives would have been. Why should I count as the "author" of my action if there was nothing in my psychological state that necessitated that action? It was just as undetermined as any of the other possible actions would have been.

Suppose, for instance, that someone decides—against his moral education but in accordance with his goals and plans—to rob a person on the street. And suppose it would have been equally possible for him to act otherwise—that is, to refrain from committing the robbery. If the outcome was genuinely undetermined, then how could the robber be responsible for it? Given agential indeterminism, the two alternative courses of action—robbery or no robbery—would have been equally possible. Shouldn't we thus treat the actual outcome as an undetermined fluke event, more or less on a par with a coin toss? It could equally have gone the other way, with the would-be robber abandoning the plan at the last minute. For the successful robber the indeterministic process went one way, while for the would-be robber who did not go ahead it would have gone the other way. The difference between these two scenarios would have been a fluke, not a matter of responsible agency.

Similarly, recall Martin Luther's choice between reaffirming his criticism of the Roman Catholic Church and renouncing his criticism. If—as I have argued—Luther's choice was genuinely undetermined, then how could his action to reaffirm his criticism count as any more "chosen" by him than the equally possible alternative of renouncing his criticism? After all, if Luther truly had alternative possibilities—as suggested by agential indeterminism—then neither action was *necessitated* by Luther's psychological state.

As should be evident, this challenge poses a threat not only for the conception of free will that I am defending here but also for any conception of free will that requires indeterminism at the point of an agent's choice. We may therefore call it the "challenge from indeterminism." Other philosophers have discussed it in relation to libertarian conceptions of free will more generally, well before the development of my own compatibilist variant of libertarianism. Robert Kane, for instance, summarizes the challenge as follows: "If an action is *undetermined* at a time *t*, then its happening rather than not happening at *t* would be a matter of *chance* or *luck*, and so it could not be a *free* and *responsible* action."[31] Responses to the challenge range from the claim that genuine responsibility is incompatible with indeterminism as much as with determinism (as, for instance, Galen Strawson argues), through the claim that what the agent is responsible for is simply the "indeterministic *effort* to decide (or choose) what to do" (as Alfred Mele suggests), to the claim that the agent can be responsible for the outcome of an indeterministic choice as long as certain conditions are met. Namely, the outcome must be *either* consistent with an already formed prior will *or* attributable to an act of "self-formation"—a process of reflection in which the agent is "torn between competing visions of what [he or she] should do" and in which one of these visions wins out, though it was not predetermined which one it would be (as Kane argues). While the first of these responses takes indeterminism to undermine responsibility, the last two responses are attempts to reconcile responsibility with indeterminism. More could be said about each response.[32]

My preferred response to the challenge, which I have developed in joint work with Wlodek Rabinowicz, makes use of the notion of intentional explanation that I have already discussed at several points in this book. Intentional explanations, as I have pointed out, rely on an important distinction between what an agent *could* do and what the agent *will* do. In particular, they usually employ some criterion for determining which actions among the possible ones count as "rational for the agent," "psychologically salient," or "singled out by certain choice heuristics," and they then predict that the agent will take an action that meets the relevant criterion. In the paradigmatic case of a rational choice, the agent chooses a *rational* course of action from among a variety of *possible* courses of action. For ease of reference, let's call the criterion for selecting an action a criterion of "intentional endorsement," though nothing much hinges on that label.

We are now able to see that the following two claims can be simultaneously true:

1. More than one course of action is *possible* for the agent, even given the agent's full psychological state.

2*. The agent's actual action is an *intentionally endorsed* one, given the agent's psychological state, whereas some (perhaps all) of the other possible actions are not endorsed.

There is no tension between these two claims at all. In fact, both are entirely in line with what an ordinary intentional explanation of the agent's action would say. For instance, our hypothetical street robber had two possible courses of action: committing the robbery, and not doing so; and it was the first of these options that he intentionally endorsed, given his overall goals and plans. He committed the robbery and did so with full intentional endorsement, though he could have acted otherwise. Similarly, Martin Luther had two possible courses of action: reaffirming his criticism of the Church, and renouncing it; and he chose the first of these options, because this was the one he intentionally endorsed, given the person he was. So, claims 1 and 2* are straightforwardly true against the background of how intentional agency works: an agent can have more than one possible course of action, and he or she then performs an action—perhaps the unique one—that he or she intentionally endorses. And if the latter claim—claim 2*—is true, then we have a good reason to reject the earlier claim 2, which asserted that the agent's actual action is just a fluke event for which he or she is not responsible. If the agent intentionally endorses the performed action, then we may plausibly attribute that action to the agent him- or herself, and not merely to chance or luck. As Wlodek Rabinowicz and I put it in an earlier article, "One can consistently say that someone who makes a choice has several alternative possibilities, and yet that, far from merely indeterministically picking an action, the agent chooses one he or she endorses."[33] This, I believe, goes at least some way towards addressing the challenge from indeterminism, over and above the challenge from determinism.[34] Of course, we would like to go further and show that the agent's action is not merely *intentionally endorsed* by him or her but also that the agent has the right kind of *control* over it. Establishing that point will be the task for the next chapter.

5

In Defence of Causal Control

Our third and final requirement for free will is causal control: The actions of any bearer of free will must be caused not just by certain non-intentional processes in this agent's brain and body, but by the appropriate mental states, specifically the agent's intentions behind those actions. If the agent had no causal control over his or her actions, then he or she would not really be in the driver's seat. The actions would be, in effect, mere occurrences over which the agent had very little influence. Indeed, we might even be reluctant to speak of "actions" in such a case, given the agent's lack of control. The agent's intentions would not be their causal source.

Think of what happens when the doctor strikes your knee with a little hammer to test your reflexes. This will prompt some involuntary movement. Would you call this a "free action"? Surely not. It is totally outside your intentional control, even though there is some bodily process that causes the movement. The mere fact that something is caused by some physical process in your body is not enough to make it a free action. You would not describe your digestive processes or your heartbeat as free actions. Rather, something can be attributed to your own free will only if it is appropriately caused by your intentions. It must be the case that you did the action *because you intended to do it*.[1]

We have seen that this causal control requirement is up against a major challenge: the challenge from epiphenomenalism. According to this challenge, there is no such thing as *mental causation*—that is, causation by an agent's intentional mental states. From a scientific perspective, so the challenge goes, anything an agent does can ultimately be attributed to nonintentional physical processes in the person's brain and body. The agent's intentions are at most byproducts of the real physical causes. They are "epiphenomena," which do not do any genuine causal work themselves. The idea that "there is some little guy in [a person's] head calling the shots," as Michael Gazzaniga puts it, is a leftover from an outdated folk-psychological way of thinking, which has no support in science. The human brain and body is a complex physical system, governed by the same kinds of nonintentional causal mechanisms that govern other physical systems. It is at the physical level that the causes of our actions are to be found, not at the level of intentional mental states. Recall Sam Harris's remarks: "Did I consciously choose coffee over tea? No. The choice was made for me by events in my brain that I, as the conscious witness of my thoughts and actions, could not inspect or influence."[2] There is no doubt that I am the *witness* of my thoughts and actions. But science seems to go against the notion that I am also their *causal source,* that I can actively control them.

My aim in this chapter is to respond to this challenge, and to defend the claim that human beings do in fact have causal control over their actions. I will explain why the so-called causal exclusion argument—the most prominent philosophical argument against mental causation—is flawed, and why the influential neuroscientific scepticism about mental causation does not quite stand up to scrutiny. The upshot is that we have good reasons to think that our actions are by and large caused by our mental states, and that it would be a mistake to attribute them solely to nonintentional physical processes.

What Do We Mean by "Cause and Effect"?

I want to begin with some preliminary remarks about the ideas of cause and effect. Causal reasoning is ubiquitous. We think about causes and effects in all realms of life. We engage in causal reasoning when we try to figure out how to light a fire to cook food; when we think about how to solve an engineering problem; when we try to treat diseases in medicine; when we look for policies to reduce unemployment or to avoid dangerous climate change; when we deliberate about whether to hold someone responsible for some harm; and when we try to answer scientific or historical questions, such as why the dinosaurs went extinct, what led to the First World War, and why there was a financial crisis in 2008. In all these cases, we are interested in *why* certain outcomes occur; what are the causes that make them happen?

Causal reasoning is central to our attempts to make sense of how the world works. It allows us to keep track of certain patterns and regularities—patterns that are often not just of theoretical interest, but that also matter for practical purposes. We not only want to explain and understand the world, but we also want to know how we can causally intervene: what we need to do in order to bring about the outcomes we want. In short, the ideas of cause and effect are relevant both to our theoretical representations of how the world works, such as in science, and to our practical reasoning in decision making and agency.

The modern understanding of cause and effect owes much to David Hume's thought in the eighteenth century. Hume was one of the giants of early modern philosophy, though he failed to gain professional academic recognition in his own lifetime due to his critical stance towards religion. He applied, but was turned down, for two distinguished academic positions: the Chair in Pneumatics and Moral Philosophy at the University of Edinburgh (note the curious combination of subjects!) and the Chair in Philosophy at the University of Glasgow. It is helpful to revisit Hume's ideas about causation briefly.

Hume's main insight is as simple as it is important: although we routinely think that there are causal relations between certain events, we never actually *observe* those causal relations themselves. We only ever observe the fact that some events are reliably followed by others. If, in conventional terms, one event "causes" another, we can certainly observe the co-occurrence of the two events—we might even see this as an instance of a repeated pattern—but we can't see the "glue" that binds these events together, over and above the fact that they reliably happen in succession. Consider, for instance, the causal claim that putting an ice cube into hot water causes it to melt. All we can observe is that whenever we put ice into hot water, it melts. That is, the event of putting ice into hot water is followed by the event of the ice melting, and the pattern seems robust. But the causal relation itself—the "necessary connection" that ties the events together—is unobservable. Hume wrote, "We have no other notion of cause and effect, but that of certain objects, which have been always conjoin'd together, and which in all past instances have been found inseparable."[3] In other words, we see the "constant conjunction" of certain kinds of events, and this leads us to think that there is a causal relation between them. But what is the status of that causal relation?

Broadly speaking, there are three different views one might take on the ontological status of causation, which correspond to different interpretations of Hume's insight. They may be called the "reductionist," "projectivist," and "realist" views about causation.[4] On a reductionist view, the lesson of Hume's insight is that the notion of cause and effect can be *reduced* to the notion of constant conjunction (which includes, for Hume, spatiotemporal contiguity and temporal succession). On this view, to say that event *A* causes event *B* is just a shorthand for saying that there is a constant conjunction between events of type *A* and events of type *B*. There is nothing more to causation than constant conjunction. On a projectivist view, by contrast, the lesson is that causal relations are projections of the human mind. Whenever we see a constant conjunction between events of type *A* and events of type *B*, we tend to *postulate* a causal relation between *A* and *B*, but this causal relation is something we project onto the world. It is a feature of our imagination, not a feature of the world. On a realist view, finally, the lesson is that causal relations are real and exist out there in the world, though they are not directly observable. On this interpretation, the constant conjunction between events of type *A* and events of type *B* is *evidence* for a causal relation, although the relation itself cannot be observed. This parallels the way in which electromagnetism, for instance, is not directly observable, though there is good indirect evidence for it. For example, we can see how metallic objects behave under the influence of what we take to be electromagnetic fields. According to a realist view, causal relations, like electromagnetic fields, are real phenomena, albeit ones we can observe only indirectly.

For present purposes, I will set aside the question of which interpretation is most faithful to Hume's own view. This is largely an exegetical question for Hume scholars. What matters for present purposes is that causal relations—even if they are real—can be observed only indirectly. As Hume has taught us, causal relations are not as visible as rocks and trees. Their identification involves an act of inference, and specifically an inference to the best explanation. We can't simply look at the world and *see* what causal relations there are. Rather, we can only *infer* causal relations from other observations, and in making those inferences we must ultimately rely on criteria such as the explanatory usefulness of the putative causal relations. Specifically, a good test for whether *A* causes *B* is whether the hypothesis that there is such a causal relation yields—or is part of—the best explanation of the patterns and regularities in which *A* and *B* stand. One aspect of this, but by no means the only one, is that the causal hypothesis might be useful in guiding our actions—for instance, by enabling us to bring about *B* via *A* or by enabling us to avoid *B* via avoiding *A*. If a causal hypothesis is explanatorily useful in the relevant sense, and it outperforms rival hypotheses, then we are warranted in accepting it, at least provisionally. Provisionally because the hypothesis might eventually be superseded by an explanatorily superior one.

So, when we are trying to answer the question of what causes what, we must ask how well a given causal hypothesis allows us to keep track of the patterns and regularities of interest and, in case practical decisions are at stake, how well it guides our actions. Does it successfully tell us which interventions in the world bring about which effects, for example? If a causal hypothesis is not explanatorily useful or practically action guiding (or at least part of a broader explanation that meets these criteria), then we don't have any good basis for accepting it. Thus, causation and explanation are intimately connected.

That said, causal relations need not be regarded as mere explanatory relations, let alone as mere practical heuristics. For all we know, causal relations may well exist out there in the world. They may well be what philosophers call "ontic" and not just "epistemic": features of the world, not just features of our cognition. But even if causal relations are real phenomena, as a realist view suggests, our best *criterion* for them may still be whether postulating them is useful from our perspective. And so, as I have pointed out, the identification of causal relations—whether in science and engineering, in medicine, or in everyday life—must always rely on an inference to the best explanation. Later I will define the notion of a causal relation a little more precisely, but for the moment the present remarks will suffice.

The Problem of Mental Causation

We can now put the problem of mental causation into sharper focus. When we are trying to find out the causes of human actions, we must ask which causal hypothesis best explains the regularities and patterns of human agency. As we have seen, explanatory considerations are ultimately our best guide to what causal relations there are. Thus, the crucial question we must ask in relation to mental causation is this: Are human actions best explained by taking people's intentions to be their causes, or are they best explained by attributing them to something else, such as certain nonintentional physical states of the underlying brains and bodies?

The defender of free will must show that a person's actions are really caused by his or her intentional mental states rather than by some subintentional physical states of the brain and body. And this is where the challenge from epiphenomenalism arises. According to that challenge, the causes of human actions are physical, not intentional: it is an illusion to think that there is any mental causation. To address this challenge, we must respond to what is arguably the strongest argument for epiphenomenalism. Unless we can rebut that argument, there is no hope of answering the present challenge. The argument is Jaegwon Kim's influential "causal exclusion argument," which I have already summarized informally in Chapter 2.[5] Later on I will revisit the neuroscientific evidence for epiphenomenalism.

Kim claims that certain fundamental principles of causation undermine the idea that mental states, as distinct from underlying physical processes, cause actions. Let me state the principles a little more precisely and explain why they seem to support epiphenomenalism.[6] Here is the first principle:

> **The causal closure of the physical world:** Any physically realized effect has a sufficient physical cause.[7]

This principle seems to be a central ingredient of a scientific worldview: there aren't any physically realized effects that lack sufficient physical causes. To think otherwise would be to accept a form of supernaturalism: some effects with a physical manifestation would be partly or even wholly caused by events outside the physical realm, without sufficient physical causes.

The second principle rules out gratuitous causal overdetermination:

> **The causal exclusion principle:** If an effect has a sufficient physical cause, then it does not have any other, distinct cause occurring at the same time.[8]

This is a principle of causal parsimony: there are no redundant causes. If some effect is wholly accounted for by some cause, then there is no other cause for the same effect, occurring at the same time.[9] Exceptions to this general rule might be, at most, some very unusual cases, such as the hypothetical case of two assassins shooting at the same human target with perfect simultaneity, where neither is uniquely causally responsible for the victim's death. We can interpret the causal exclusion principle as not applying to such cases. We may call them "cases of genuine causal overdetermination."[10]

Why is the causal exclusion principle plausible? Recall what I have said about causal reasoning involving an inference to the best explanation. Clearly, if we have identified a cause that fully accounts for some effect, then it would normally be redundant to postulate a second, simultaneous cause that allegedly also accounts for that effect. It would not be supported by an inference to the best explanation (except perhaps in some rare cases of genuine causal overdetermination). We should not gratuitously overascribe causation.

We can now see the implications of these two principles. The causal closure principle implies that, insofar as any human action manifests itself in some physical event, and it has a cause at all, it must have a sufficient physical cause. Presumably this will be a cause within the person's brain and body. It would be spooky to suggest that my hand movement, despite having a cause, lacks a sufficient *physical* cause. But now the causal exclusion principle kicks in. If the action has a sufficient physical cause, then it does not have any other, distinct cause occurring at the same time.[11] So, contrary to what we conventionally think, the relevant person's intention—the appropriate mental state—cannot qualify as a cause of the action. This, then, seems to rule out mental causation.

Or, to be precise, it rules out mental causation *unless* we identify the agent's mental states with the underlying physical states of the brain and body. To *identify* the agent's mental states with the underlying physical states would be to assume that the mental states are *nothing over and above* those physical states. If we identified the agent's intention with some physical state—say, a particular brain state—then the intention would trivially qualify as the cause of the agent's action. The purported "mental cause" would be nothing over and above the relevant physical cause. However, as I have argued at length, intentional mental states are not reducible to any physical states of the brain and body. Specifically, I have defended the following principle:

> **The nonidentity principle:** An agent's mental states are not identical to any underlying physical states of the brain and body, even though they supervene on physical states.

And so, there is no way of rescuing mental causation simply by claiming that what we conventionally call "the agent's intention" is just some complicated physical state of the brain and body, where that physical state is the one that causes the agent's action.

Recall that to say that the agent's mental states "supervene" on underlying physical states is to say that those mental states are a necessary consequence of the physical states in question: whenever those physical states obtain, then so do the resulting mental states. This supervenience thesis is a central tenet of a scientific worldview. But crucially, as we have also seen, the supervenience thesis must be supplemented with a nonreductive qualification: even though mental states supervene on underlying physical states, they are not reducible to those physical states. This is (among other things) because their specification involves intentional properties—such as beliefs, desires, and intentions—which are beyond the conceptual repertoire of physics. Mental states have a paradigmatic higher-level character, not a physical one. Remember the ugly but descriptive word "aboutness." Mental states and mental properties have "aboutness." They are about something; they have meaningful contents. Physical states and physical properties do not have any "aboutness." They are not "about" anything; they have no meaningful contents. The relations in which they stand are of a purely physical sort. For this reason, it would be a category mistake to think of a person's beliefs, desires, and intentions as being nothing more than physical states or physical properties of the person's brain and body.

To sum up, the causal closure and exclusion principles jointly imply that there is no such thing as mental causation. We would only be able to avoid this conclusion if we were to give up the nonidentity principle. But this principle is hard to give up unless we are prepared to deny the higher-level character of mental states. And as we have seen, this would be an implausible move.

Note that this theoretical argument presents a principled obstacle for mental causation. If the argument is correct, then the neuroscientific evidence that I have reviewed earlier, such as Benjamin Libet's series of experiments (see Chapter 2), is not even strictly needed to settle the matter. From the perspective of the causal exclusion argument, it is no surprise that the cause of a subject's action in a psychological experiment, such as his or her hand movement, is not a conscious intention but some underlying physical, neural state. Libet's experiments, it appears, merely confirm what theoretical considerations imply already—namely, that we do not have the intentional control over our actions that we conventionally think we have.

So, where do we go from here? Is free will doomed on the grounds that there is no such thing as mental causation, even in principle?[12]

The Causal Exclusion Argument Generalized

The causal exclusion argument conveys a picture of causation according to which causal relations are ultimately confined to the physical level. On this picture, there are no genuine causal relations at the mental or psychological level, because they would be "excluded" by underlying causal relations at the physical level. To suggest that there could be mental causation would go against either the causal closure principle or the causal exclusion principle.

It is worth noting that the logic of this argument generalizes. It carries over to any phenomenon that can be considered at lower and higher levels of description, whether it is the phenomenon of human agency, on which I have focused here, where the competing levels are the physical and psychological ones; the phenomenon of an organism or ecosystem, where the competing levels are those of molecular biology and those of systems biology; or any social or economic phenomenon, where the competing levels are the individual and aggregate ones. In each of these cases, the logic of the exclusion argument suggests that causal relations must ultimately be identified at the lower, more microscopic level of description, not at the higher, more macroscopic one.

In particular, this conclusion—that there can only be lower-level causation—follows from two principles that are straightforward generalizations of the causal closure and exclusion principles.[13] The first is a "lower-level causal closure principle," which asserts that any purported higher-level cause for some effect must be underwritten by some sufficient lower-level cause. There must be lower-level causal foundations: we must not postulate free-standing higher-level causes without underlying lower-level causes. And the second principle is a "generalized causal exclusion principle," which asserts that if an effect has a sufficient lower-level cause, then it does not have any additional higher-level cause occurring at the same time; to suggest that it did would once again amount to a form of gratuitous causal overascription. The two principles immediately imply that all causal relations must be found at the lowest level under consideration; lower-level causes exclude higher-level ones.

Let's consider for a moment what follows from this line of reasoning. In the case of human behaviour, as we have already seen, all causal work must take place at the physical level, not the mental one. In the case of biology, all causal work must take place at the cellular or molecular level, not at the level of organisms or larger systems. And in the case of social phenomena, all causal work must take place at the level of individuals, not at the level of larger social structures.

For instance, when we say that the decrease in the population of the great predators causes an increase in the population of their prey and, as a further consequence, a depletion of their food resources, this claim would be, at best, a shorthand for a more complicated causal hypothesis at the level of the individual animals or—even better—at the level of their cellular components. Similarly, it would be incorrect to claim that there are genuine causal relations in macroeconomics, such as when we say that an increase in the interest rate causes a decrease in inflation. What's really going on, according to the generalized causal exclusion argument, is that the individual actions of a large number of market participants cause some aggregate outcome, which amounts to a decrease in inflation.

Is this picture of causation plausible? Is causation really confined to the lowest level of description, so that there can never be any genuine higher-level causal relations?

Why It Would Be a Mistake to Think That Causation Is Confined to the Lowest Level

In what follows, I will argue that there are at least three major reasons why we should resist the view that causation is confined to the lowest level, such as the microphysical level or similar. The first reason is that if we really accepted that view, this would have a rather dramatic implication, even more dramatic than the implication that there cannot be any higher-level causation.[14] The implication is this:

> **Problem 1:** Unless there exists a "fundamental level of reality," causation would run the risk of "draining away" altogether. There might not be any causal relations at all, whether higher- or lower-level ones.

Why is this? Imagine—for the sake of argument—that, for every level at which we might describe the world, there is, in principle, an even lower, more basic level. So, no matter how deep we dig in our scientific investigations, we never hit "rock bottom." Then, according to the causal exclusion argument, any causal hypothesis—however putatively fundamental—will be subject to a challenge from epiphenomenalism, because the real causal work would have to take place at an even lower level. In an article titled "Do Causal Powers Drain Away?," Ned Block makes this point:

> If there is no bottom level, and if every (putatively) causally efficacious property is supervenient on a lower "level" property (Call it: "endless subvenience"), then (arguably) Kim's Causal Exclusion Argument would show . . . that any claim to causal efficacy of properties is undermined by a claim of a lower level, and thus that there is no causation.[15]

Now, of course, we can avoid this conclusion by insisting that there is a bottom level of reality, at which all genuine causal relations are located. As Kim himself points out, "It is generally thought that there is a bottom level, one consisting of whatever microphysics is going to tell us are the most basic physical particles out of which all matter is composed (electrons, neutrons, quarks, or whatever)."[16] If this is right, causal relations at the microphysical level can be said to be truly real, and we can recognize causation as a fundamental physical phenomenon. Indeed, some scientists and philosophers might welcome this picture of causation. Perhaps this is what a robust materialist or physicalist worldview comes down to.

But, contrary to this picture, we cannot take for granted that there is such a thing as a "fundamental level of reality."[17] From a scientific perspective, it remains an open question whether there is a bottom level. If we look at the history of science, the human understanding of the world has become ever more fine grained over the years, and this trend has not yet stopped. Each time people thought they had found the fundamental level, new developments eventually led to an even more fine-grained understanding of the world. People once thought that atoms were indivisible, only to learn later that atoms actually consist of protons, neutrons, and electrons. That understanding was subsequently refined further with the discovery of even smaller constituents, such as quarks, leptons, and bosons. Physicists are now discussing whether the world might be made up of tiny vibrating strings, the fundamental building blocks according to string theory. Yet it would not be surprising if this were not the last word on this issue either. Given that history, it is not so far-fetched to speculate that there might be no bottom level at all. However deep we dig, we might always discover even finer levels of reality which we had previously overlooked. I am not taking a stand on this possibility here, except to agree with Block's remark that "the hypothesis that there is no bottom level . . . appears to be an open question, not a mere philosopher's possibility like the possibility that the world was created 5 seconds ago complete with the evidence of an ancient provenance."[18] The worry, then, is that the causal exclusion argument renders the very idea of causation contingent on a controversial hypothesis—namely, that there is a bottom level at which all causal relations are located. If we cannot assume that there is such a level, then, given the logic of the exclusion argument, the entire notion of causation is in trouble. What began as an argument against *higher-level* causation has turned into an argument against causation in all forms. Of course, sceptics about causation might be prepared to bite the bullet on this conclusion. But to me, it looks more like a reductio ad absurdum of the causal exclusion argument itself.

The problems for the view that causation is confined to the lowest level do not end here. The second reason why we should resist that view is this:

> **Problem 2:** Even if there is a most fundamental level, such as the one described by our best theories of physics, it is not at all clear that this level is hospitable to causation as we normally understand it.

In particular, the ideas of cause and effect do not seem to have much of a place in current fundamental physics. In the sciences, cause and effect reasoning is much more common in the special sciences, such as the biomedical, human, and social sciences, than in fundamental physics, like classical and quantum mechanics. The causal exclusion argument would therefore prompt us to look for causal relations at a level at which such relations are least likely to be found. To explain this point, let me elaborate a bit.

It is helpful to go back to the early twentieth century. In 1913 Bertrand Russell published an essay titled "On the Notion of Cause," in which he wrote, "The law of causality . . . like much that passes muster among philosophers, is a relic of a bygone age, surviving like the monarchy, only because it is erroneously supposed to do no harm."[19] Specifically, Russell argued, the traditional notions of cause and effect are out of sync with modern science. Russell was referring to science as it was practiced at his time, but his points remain of interest today. What was Russell's concern? Causation, as traditionally understood, is a discrete relation between distinct events, such as the event of putting an ice cube into water and the event of the ice melting, or the event of falling down the stairs and the event of injuring oneself. Fundamental physics, however, does not carve up the world in this way. It neither employs an ontology of "events" nor posits discrete relations of the traditional cause and effect kind. Rather, it seeks to identify general laws governing physical systems in their entirety and focuses on continuous functional relations between continuous variables.

As an illustration, consider Newtonian mechanics. This describes the world as consisting of point particles. The physical state of each particle at a particular time is given by its position and momentum. If a system consists of several particles, then the system's state is given by the positions and momenta of all its constituent particles. The system's state space consists of all the possible states in which the system could be. Newton's equation of motion determines what state transitions are allowed; for instance, whether the system could be in state y at time $t=1$ if it was in state x at time $t=0$. In this way, we can infer the histories of the system that are consistent with Newton's laws (the trajectories the system can take through its state space). Under normal circumstances, as already noted, a Newtonian system is deterministic: the system's state at any particular time determines all future states. This description of the world does not recognize discrete events, and so there is nothing in it that could be labeled as either "cause" or "effect." In Russell's words,

> In the motions of mutually gravitating bodies, there is nothing that can be called a cause, and nothing that can be called an effect; there is merely a formula. Certain differential equations can be found, which hold at every instant for every particle of the system, and which, given the configuration and velocities at one instant, or the configurations at two instants, render the configuration at any other earlier or later instant theoretically calculable. . . . But there is nothing that could be properly called "cause" and nothing that could be properly called "effect" in such a system.[20]

Of course, we could still use the term "causal laws" to refer to the general principles and constraints governing a physical system. However, those "causal laws" have very little in common with the discrete cause and effect relations postulated in traditional causal reasoning. The language of "causation" seems entirely optional in modern fundamental physics. Instead of referring to "causal laws," we could equally just speak of the "laws of nature."

Russell's view that the notion of cause and effect has no place in fundamental physics is echoed in more recent debates. In describing the mainstream view, James Woodward, a leading philosopher of causation, notes that "the view that fundamental physics is not a hospitable context for causation and that attempts to interpret fundamental physical theories in causal terms are unmotivated, misguided, and likely to breed confusion is probably the dominant, although by no means universal, view among contemporary philosophers of physics."[21] Woodward himself is a proponent of a theory of causation that is well suited for the special sciences—the interventionist theory, on which I will say more later.

Modern physics raises some further difficulties for the traditional understanding of causation, in addition to those identified by Russell. On the traditional understanding, causation is asymmetrical in time: causes precede their effects. However, since Albert Einstein devised the theories of relativity, the notion of temporal succession—the order in which things happen—is no longer taken to be absolute, but only relative to a reference frame. Whether one event precedes or succeeds another may depend on the frame of reference adopted by the observer. If I am stationary on Earth, I may perceive event *A* as happening before event *B*. If, instead, I am travelling in a spaceship at very high speed, perhaps close to the speed of light, I may observe these two events in the reverse order—that is, *B* before *A*. Crucially, the point is not merely that I receive the *evidence* about these two events in a different order, depending on my reference frame. Rather, time itself is relative to the reference frame. This may seem a little mind-boggling, but it does raise significant questions for the idea that causes precede their effects. If the temporal order is not absolute, but only relative to the reference frame, then, it seems, causal relations themselves can no longer be absolute, but only relative to a reference frame.

In light of all these points, it seems reasonable to conclude that traditional notions of cause and effect are not particularly well suited for *fundamental* physics.[22] Thus, arguably, the view that causation is confined to the lowest level misidentifies the most plausible place of causation in the world.

This takes us straight to the third and final reason why we should reject that view:

> **Problem 3:** If we insisted that there are no higher-level causal relations, we would fail to do justice to the fact that cause and effect reasoning is indispensable in many special sciences, such as the biological, medical, human, and social sciences.

Think of what we are interested in when we study higher-level phenomena such as biomedical, social, and economic phenomena. We want to know *why* things happen, and in answering that question we want to isolate the factors that make a difference to the outcomes in question. This in turn is relevant to a second question: *How* can we bring about certain outcomes that we want to achieve, and avoid other outcomes that we do not like? Here are some familiar examples of the "why" and "how" questions in the special sciences: Why is there a loss of biodiversity, and how can we stop it? Why do some people get Alzheimer's disease, and how can we medically treat them? Why do financial crises happen, and how can we prevent them? Why are some democracies stable while others collapse, and how can we improve democratic stability? And why does a person act in one way rather than another—say, he or she commits a crime—and how can certain bad behaviours be successfully avoided?

To answer these and many other special-science questions, we need to engage in causal reasoning. Merely documenting correlations or non-causal patterns is not enough. Similarly, pointing to the fundamental laws of nature governing physical systems in their entirety is not of much help either. Nancy Cartwright made this point forcefully in a classic article in the 1970s, and it has been widely recognized since then: "abandoning the concept of causation would cripple science," in Hartry Field's words.[23] Let me explain why cause and effect reasoning is indispensable in the special sciences.

Consider the "why" and "how" questions I have mentioned. To answer them, we need to figure out not only which events or phenomena are *associated* with which others but also which events or phenomena *make a difference* to which others. A frequently cited example is the following: Smoking is associated with lung cancer and also with yellow teeth. Having yellow teeth, by being associated with smoking, is also associated with lung cancer. But *why* does someone get lung cancer? Is it because they smoke, or because they have yellow teeth, or for some other reason? Clearly, if we wish to reduce the incidence of lung cancer in the population, it is not enough to know what other things smoking is correlated with. It would be bad advice simply to send people to the dentist for dental cleaning or to ask them to use whitening toothpaste. Changing the colour of a person's teeth makes no difference to his or her lung-cancer risk. If the person carries on smoking, he or she is just as likely to get lung cancer as before, no matter how shiny his or her teeth are. By contrast, getting the person to stop smoking makes a big difference. This is because the relationship between smoking and lung cancer is a *causal* one, while that between yellow teeth and lung cancer is a mere *correlation*.

Cartwright distinguishes between what she calls "laws of association" and "causal laws." Laws of association are the kinds of laws that Russell took to be at the heart of the physical sciences. These describe which variables or quantities are *systematically associated* with which others, but without specifying any explicit cause and effect relationships. Causal laws, by contrast, tell us which variables or quantities *make a difference* to which others, so that by changing the "cause variables" we can systematically bring about changes in the "effect variables." Cartwright's point is that, to answer the "why" and "how" questions mentioned above, identifying mere laws of association is not enough. We need causal laws. These, in turn, yield what Cartwright calls "effective strategies" for intervening in the world.

In line with these observations, it has become increasingly common to understand causation in terms of "difference making" or "intervention." When we engage in causal reasoning, we want to know which actual or hypothetical interventions in the world would make a difference to which effects. I will use the term "causation as difference making" to refer to this idea. The idea is a common denominator of several recent theories of causation in philosophy, statistics, and related fields, including so-called counterfactual, probabilistic, interventionist, manipulationist, and contrastive theories (associated with scholars such as David Lewis, Judea Pearl, Joseph Halpern, James Woodward, Peter Menzies, and Huw Price).[24]

An understanding of causation as difference making is central to the special sciences. Much of the methodology of inferential statistics, for example, revolves around the goal of determining which variables make a significant difference to which others, while holding fixed certain control variables. The same is true of the methodology of randomized controlled experiments. Using such methods, we have been able to find out that regularly taking a low dose of aspirin reduces the heart attack risk in certain kinds of patients, while eating lots of butter does not. We have learnt that air pollution makes a difference to cancer rates in the population, while the use of microwave ovens (as far as we know) does not. And we have established that increased carbon dioxide (CO_2) concentrations in the atmosphere make a difference to the average global temperature, while increased concentrations of, for instance, nitrogen (N_2) do not. In each of these cases, scientists identify causation by looking for evidence of difference making, especially difference making that persists even when we control for other factors. Causes are difference makers, while anything that is not a difference maker is not a cause. It is hard to imagine what the special sciences would look like if we didn't have such reasoning at our disposal.

The foregoing considerations should be enough to make the case for resisting the view that causation is confined to the lowest level. The denial of any higher-level causal relations would seriously hamper the special sciences. But to establish that there truly is higher-level causation, it is not sufficient to show that its denial has undesirable consequences. We must also convince ourselves that the causal exclusion argument is unsound. Unless we can refute some of the argument's premises, we should be uneasy about rejecting its conclusion. It is this challenge to which I will now turn.

The Case for Realism about Causal Control

My aim in this section is to give a positive argument for the existence of higher-level causation in general and mental causation in particular, and to explain where the causal exclusion argument goes wrong. My analysis is based on joint work I have done with Peter Menzies.[25] Others, including James Woodward, Panu Raatikainen, John Campbell, Adina Roskies, and Jenann Ismael, have come to similar conclusions.[26]

As I have noted, from a special-science perspective, causation is best understood as "difference making." So, when we are looking for the causes of human actions or any other special-science phenomena, we must ask which factors make a systematic difference to the effects in question. For example, in the case of human action, we must identify the neural, psychological, or other properties which make it the case that a person acts in one way rather than another. Or, in the case of economic and social phenomena, we must identify which variables make a difference to the economic or social outcomes of interest, such as the inflation rate, the crime rate, or the support for extremist political parties.

I will now show that when we understand causation in this way, we are warranted in concluding that higher-level effects often have higher-level causes. Contrary to the causal exclusion argument, causation is often to be found at the higher level alone. Let me begin by giving a simple, but useful definition of causation as difference making. A putative cause, *C*, *makes a difference* to a putative effect, *E*, if and only if the following two conditions hold:

1. If *C* were to occur, then *E* would occur.
2. If *C* were not to occur, then *E* would not occur.

Let us call the first of these "the positive conditional" and the second "the negative conditional." Remember that a statement of the form "if *X* were the case, then *Y* would be the case" stands for the claim that in the nearest possible worlds in which *X* is true, *Y* is also true. So, to determine whether *C* makes a difference to *E*, we must ask

- whether in the nearest possible worlds in which *C* occurs, *E* occurs as well; and
- whether in the nearest possible worlds in which *C* does not occur, *E* does not occur either.

For example, falling down the stairs causes me to get injured. This is true because

- in the nearest possible worlds in which I fall down the stairs, I get injured; and
- in the nearest possible worlds in which I do not fall down the stairs, I do not get injured.

Of course, there are many technicalities that need to be addressed if we wish to use this criterion of causation in practice. For instance, we need to specify what we mean by "the nearest possible worlds." We need to explain how to generalize the present criterion for difference making from the simple cases where causes and effects are binary (they occur or do not occur) to more complicated cases where they are continuous (think of the interest rate or the inflation rate). And we must explain how we can control for other, potentially confounding factors in our analysis of difference making. After all, *C makes a difference* to *E* only if the regularity in which *C* and *E* stand is not driven by some common cause that is responsible for both. For present purposes, I will set these complications aside, and refer the reader to the literature on causal modelling.

The point I want to make is that, in the case of many higher-level effects, the positive and negative conditionals for difference making are satisfied *only* by appropriate higher-level causes, not by lower-level ones. Let me begin with some illustrations distinct from mental causation. The first example—borrowed from Frank Jackson and Philip Pettit—is a macroscopic physical one. Suppose you heat up a flask of water, and when the water boils, the flask breaks due to the pressure.[27] What is the cause of the flask's breaking? The most natural answer is that it is the boiling of the water, which increases the pressure beyond the level the flask can withstand. And indeed, this answer is confirmed by our analysis of causation as difference making. If we take *C* to be "the boiling of the water" and *E* to be "the breaking of the flask," then the positive and negative conditionals for difference making are true:

- in the nearest possible worlds in which the boiling occurs, the breaking of the flask occurs as well; and
- in the nearest possible worlds in which the boiling does not occur, the breaking of the flask does not occur either.

Suppose, however, that we insisted that the cause of the flask's breaking is not the boiling of the water but the underlying microstate of the millions of water molecules. So, *C* would be the molecular microstate that realizes the boiling on a given occasion, rather than the boiling itself as the supervenient higher-level event. But then the two conditionals for difference making would no longer be satisfied. Although it would still be true that

- in the nearest possible worlds in which the given microstate occurs, the breaking of the flask occurs as well,

it would no longer be true that

- in the nearest possible worlds in which that microstate does not occur, the breaking of the flask does not occur either.

The reason is that the boiling of the water is multiply realizable: many different microstates can realize the boiling. And many such microstates would equally make the flask break. So, the negative conditional for difference making would come out false. In particular, in some of the nearest possible worlds in which the given microstate does not occur, the breaking of the flask would still occur.

A second example comes from macroeconomics. The central bank raises the interest rate, and inflation goes down. What is the cause of the decrease in inflation? The standard answer is that it is the increase in the interest rate. And in line with the present analysis, the increase in the interest rate satisfies the positive and negative conditionals for difference making:

- in the nearest possible worlds in which the increase in the interest rate occurs (other things being equal), the decrease in inflation occurs as well; and
- in the nearest possible worlds in which the increase in the interest rate does not occur (again, other things being equal), the decrease in inflation does not occur either.

By contrast, suppose we were to insist that the decrease in inflation should be attributed to the detailed voting pattern of the monetary policy committee members at the central bank, or to their bodily states. Or alternatively, suppose we were to enumerate the actions of a very large number of market participants whose combined effect is the decrease in inflation. In each of these cases, we would have failed to identify the difference-making cause of the decrease in inflation. The way in which the interest rate increase is brought about and the mechanism by which inflation goes down are multiply realizable. Many different voting patterns within the central bank's monetary policy committee can instantiate the same interest rate decision, and many different sets of market transactions can lead to the same decrease in inflation. So, it is not true that

- in the nearest possible worlds in which the individual-level actions are a little different while the increase in the interest rate still occurs, the decrease in inflation does not occur.

We can express this point by saying that the decrease in inflation is robust to changes in the microlevel backstory which produces that effect, just as the breaking of the flask in the earlier example is robust to changes in the molecular microstate realizing the boiling of the water. In both examples, the lower-level realizers of the relevant higher-level causes do not themselves meet the conditions for difference making.

Now let's turn to the case of mental causation. Suppose I hail a taxi somewhere in London, and I ask the driver to take me to St Pancras Station. Predictably, I will be driven to that destination. On another day, I hail another taxi, but this time I ask the driver to take me to Paddington Station. Again, I will be driven to the requested destination. What is the cause of the driver's taking me to one place rather than another in each of these cases? What, for example, causally explains the fact that the driver takes me to St Pancras on the first day, and to Paddington on the second? Someone preoccupied with the idea of physical-level causation might look for causes in the microphysical or biochemical state of the driver's brain and body, or perhaps (though rather misguidedly) in the microphysical state of the car. But as in the examples of the flask and inflation, this would be the wrong level at which to look for a cause. By focusing on microphysical details, we would fail to identify the main factor that makes the difference to the taxi's destination—namely, the driver's intention: a relevant mental state.

On day one, when I say "To St Pancras, please!," the driver forms the intention to take me to St Pancras and has a rational incentive to do so. On day two, when I say "To Paddington, please!," the driver forms a different intention—this time to take me to Paddington—and once again has an incentive to do so. The microphysical details of the driver's brain and body, not to mention those of the car, are irrelevant from a difference-making perspective. The difference-making cause of the driver's action is his or her intention to take me to one place rather than another. If we take *C* to be "the driver's intention to take me to St Pancras" and *E* to be "the driver's action of taking me to St Pancras," then the positive and negative conditionals for difference making are clearly satisfied:

- in the nearest possible worlds in which *C* occurs, *E* occurs as well; and
- in the nearest possible worlds in which *C* does not occur, *E* does not occur either.

By contrast, if we substituted the detailed microstate of the driver's brain and body for *C*, the negative one of these conditionals would no longer be satisfied. If the driver's bodily or neural microstate were a little different, he or she would still take me to St Pancras, holding fixed the intention. As noted before, intentions are multiply realizable: different bodily and neural microstates of the driver could still realize the same intention and lead to the same overall effect. Generally, microphysical brain states are too fine grained to serve as the difference-making causes of human actions. They encode extraneous details that make no difference to the resulting actions. By contrast, intentions do make a difference.

In each of the examples I have discussed, the difference-making cause of the effect under consideration is to be found at the higher level rather than the lower one. The breaking of the flask is caused by the boiling of the water, not the underlying microstate of the water molecules. The decrease in inflation is caused by the increase in the interest rate, not by the individual-level actions realizing this increase. And the driver's action of taking me to one place rather than another is caused by the driver's intention, not by the physical microstate of his brain and body. The causal relation in each of these examples is what we call "realization insensitive": the effect in question would continue to occur under certain perturbations in the way the cause is realized at a lower level.[28] Realization insensitivity typically occurs when higher-level phenomena, as well as the regularities in which they stand, are multiply realizable. The boiling of the water, the increase in the interest rate, and the taxi driver's intention each admit a variety of lower-level realizing conditions, and irrespective of which of these obtain, the resulting higher-level regularities are largely unaffected. When a higher-level difference-making relation is realization insensitive, it is not accompanied by a matching difference-making relation at the lower level.

To see that these observations matter, think of how we intervene in the world when we seek to bring about a desired effect or avoid an outcome we do not like. To avoid the breaking of the flask, for example, we regulate the temperature of the water so as to prevent it from boiling. We do not simply perturb the molecular microstate of the water; indeed, this would not generally be sufficient to prevent the breaking of the flask. Similarly, when the central bank wishes to reduce inflation, it increases the interest rate; it does not merely target the underlying microlevel details. And finally, when we seek to intervene in matters related to human agency, we engage with the relevant people at the intentional level, not the neural one. For instance, we ask the taxi driver to take us to one destination rather than another; we do not simply "manipulate" his or her brain. In fact, manipulating people's brains would be the wrong way to engage with people, from a practical perspective as much as from an ethical one. In everyday life we influence one another's actions by changing one another's mental states, often through communication.

As Adina Roskies has argued, the control variables of many higher-level phenomena, including human actions, are themselves located at a higher level. Drawing on the work of John Campbell, she defines *control variables* as "parameters which, when changed, lead to systematic changes in other variables of interest."[29] Roskies emphasizes two important features of such variables: "1) control variables do not exhibit 'gratuitous redundancy,' and 2) they are 'manipulable by local processes.'"[30] In relation to human action, she endorses the following passage from Campbell:

> (a) psychological variables [capturing intentional mental states] function as control variables for the outcomes in which we are interested, (b) what is going on at a psychological level of description supervenes on what is going on at a physical level of description, but (c) at the physical level, there are no control variables for the outcomes in which we are interested.[31]

This nicely summarizes my present conclusion: It is a person's intentional mental states that are normally the difference-making causes of the person's actions, not the underlying physical states of the brain and body. And this is entirely compatible with recognizing that mental states are physically realized at the level of the brain. It is just that the realizing brain states do not themselves qualify as difference-making causes of the resulting actions. Note that, in special cases in which a person's actions are not under the causal control of any intentional mental states—as, for instance, in the case of certain neurological disorders or other subintentional difference makers—we would not normally say that those actions are freely performed.

We can now see what has gone wrong with the causal exclusion argument. Although the argument does not refer to causation as difference making, its two main premises—the causal closure principle and the causal exclusion principle—lose their force once we take on board what we have learnt about difference-making causation. Remember that the first premise—the causal closure principle—asserts that any higher-level effect has a sufficient lower-level cause. And the second premise—the exclusion principle—asserts that any effect with a sufficient lower-level cause does not also have a distinct, higher-level cause occurring at the same time. Note the reference to "sufficient cause" in each of these principles. As I will now explain, depending on how exactly we disambiguate that notion, one or the other of these principles is false.

Suppose that by "sufficient cause" we simply mean a state of affairs such that *if* that state of affairs were to occur, *then* the relevant effect would (of necessity) occur as well. On this interpretation, "sufficient cause" means something like "nomologically sufficient condition": a condition that is sufficient for the occurrence of the effect under the laws of nature. Then the lower-level causal closure principle is certainly true. After all, any higher-level effect, by supervening on certain lower-level phenomena, will undoubtedly be the consequence of certain things that happen at the lower level, and will therefore have a nomologically sufficient condition at that level. In the example of the breaking flask, it is the molecular microstate of the boiling water. However, the causal exclusion principle will be false if we interpret "sufficient cause" in this way. As my examples have shown, the existence of a nomologically sufficient lower-level condition, such as the microstate of the water, is entirely compatible with the existence of a higher-level difference-making cause, such as the boiling itself. In fact, the boiling is not only a difference-making cause of the breaking; it also qualifies as a nomologically sufficient condition for it, albeit a macroscopic one. So, even if we ignore difference-making causes altogether, we cannot say that the molecular microstate is the only candidate for a nomologically sufficient condition for the breaking. Thus the exclusion principle is false under the present interpretation of "sufficient cause."

Suppose, on the other hand, that by "sufficient cause" we mean a difference-making cause. Then, even if the causal exclusion principle were true, the lower-level causal closure principle would be false. The reason is that, while any higher-level effect can certainly be expected to have a *nomologically sufficient condition* at the lower level, it need not generally have a lower-level *difference-making cause.* As we have seen, whenever a higher-level difference-making relation is realization insensitive, it is not accompanied by a corresponding difference-making relation at the lower level. In all the examples I have considered—the breaking flask, inflation, and the taxi driver—the difference-making causes are to be found exclusively at the higher level, not at the lower one. This is despite the existence of nomologically sufficient lower-level conditions. So, the lower level is not causally closed if causation is understood as difference making. The causal exclusion principle, on the other hand, happens to be satisfied, but in the opposite way to the one envisaged by Kim. It is a difference-making cause at the higher level that seems to "exclude" any difference-making cause at the lower level, not the other way around. I should note, however, that the causal exclusion principle may itself fail too, insofar as higher-level causal relations need not always be realization insensitive. It is theoretically possible that higher-level causal relations are realization sensitive, in which case higher-level and lower-level difference-making relations can coexist. For our present purposes, I set this complication aside.[32]

In sum, the causal exclusion argument is unsound.[33] There can be higher-level causation which is not accompanied by any matching lower-level causation. Mental causation is an important instance of this. And so, we have every reason to think that a person's causal control over his or her actions is a real phenomenon.

The Libet Experiments Revisited

Before concluding, I must revisit the Libet experiments, which are often invoked to challenge the reality of mental causation. As already noted, Libet conducted a series of experiments in the 1980s that purportedly show that certain voluntary actions, such as spontaneous hand or finger movements, are caused by subconscious physical states of the brain and not by the conscious intentions to perform those actions.[34] In Libet's experiments, since then replicated by others, each subject was asked to perform a spontaneous movement, such as a flexion of the fingers or wrist, at a time of his or her choice. Subjects faced an easily readable clock, which enabled them to report the time at which they formed (or became aware of) the intention to perform the movement. Libet further measured the electrical activity in each subject's brain via an electroencephalogram (EEG). This allowed him to compare three points in time:

1. The time of the conscious intention, as reported by the subject.
2. The time of the onset of a neuronal readiness potential in preparation for the action, as measured by the EEG.
3. The time of the action itself.

On a naive, prescientific understanding of free will, one might expect that the formation of the conscious intention precedes the onset of the neuronal readiness potential, which in turn precedes the action. Libet discovered, however, that the temporal order was very different: the onset of the neuronal readiness potential *precedes,* rather than *succeeds,* the subject's conscious intention to act. As Patrick Haggard summarizes the finding, "When subjects use a clock hand to estimate the time at which they first experienced the conscious intention that led to a voluntary action, conscious awareness of intention lags the onset of RP [the neuronal readiness potential], raising a challenge for the traditional Cartesian concept of conscious free will."[35] Does this mean that the subject's action cannot be attributed to the intention, but that it was the brain that made him or her do it?

Much has been written about what Libet's experiments do or do not show, and I have little to add to this already very rich debate.[36] Here I simply want to put the experiments into perspective in light of my discussion of causation as difference making. Specifically, I want to suggest that, even if we grant Libet's empirical observations, they do not undermine the status of intentions as the difference-making causes of human actions. It would be a mistake to interpret the observed neuronal readiness potentials as the difference-making causes, even though those readiness potentials undoubtedly play an important role among the *lower-level realizing conditions* of human agency.

For a start, let's take a closer look at Libet's finding that the onset of a subject's neuronal readiness potential precedes, rather than succeeds, the subject's conscious intention to act. Does this challenge the view that the action can be causally attributed to the subject *qua* agent? Note that, on a scientific worldview, it is beyond dispute that the subject's agency is ultimately the result of physical processes in the brain and body. Even if the phenomenon of agency cannot be adequately *explained* in physical terms, it still *supervenes* on physical processes. Any conscious intention must therefore have some physical realization. It would be very mysterious if someone could form a conscious intention without any trace of neural activity in his or her brain. We would consider this a breach of the laws of nature. The conscious intention, which is a higher-level property, could not occur without a lower-level realizer, and for this reason we should not expect to find the conscious intention *before* any relevant neuronal activity.

Now, admittedly, it would be theoretically neat if the onset of the neuronal readiness potential *perfectly accompanied* the appearance of the subject's conscious intention. Libet showed that it *precedes* it. But, on reflection, the fact that there is a time lag is not very surprising. We must not confuse the *formation of the intention* with the *conscious awareness* of it. Presumably, the formation of an intention takes a little while, and, as in the case of other extended processes, we should not be surprised that the onset of the process slightly precedes the subject's conscious awareness of it. As Alfred Mele asks, "If informed conscious reasoning leads to a decision that is then put into action, why should it matter if there is a bit of a lag—a couple hundred milliseconds—between when the decision is made and when a person becomes conscious of it?"[37] Once we have freed ourselves from the initial pretheoretic sense of surprise, we can view Libet's experiments as giving us insights into some of the lower-level realizing mechanisms of human decision making without challenging the role played by higher-level intentions.

Second, as Mele has also argued, there is nothing in Libet's experiments that prevents us from interpreting the onset of the neuronal readiness potential as marking the beginning of the subject's decision process, rather than as marking the completed decision itself. Mele writes, "Maybe what's going on in the brain when the rise [in the readiness potential] begins is a process that might—or might not—lead to a decision a bit later."[38] Consider the analogy of a committee decision in an organization. A hiring committee, for example, decides whether to give a job to a particular applicant. Here, the decision maker is the committee as a whole (a "group agent"). No doubt there will be all sorts of "lower-level" preparations in the run-up to the decision. The committee members will each read the application materials, and they might even prepare a draft of an offer letter, so as to ensure a prompt implementation of their decision once it is made. A sociologist studying this process may well interpret this preparatory activity as a kind of "readiness potential" for the resulting committee decision, where the onset of the readiness potential can be quite early. Still, the decision itself takes place only when the committee officially meets and takes a vote. The presence of a readiness potential preceding the actual decision does not undermine the fact that the committee itself has control over the resulting outcome. The vote of the committee satisfies the positive and negative conditionals for difference making.

Going back to Libet's experiments, my suggestion that the observed neuronal readiness potentials should not be interpreted as the difference-making causes of the subjects' actions is further supported by an observation made by Libet himself. In his classic study, Libet acknowledges that, even after the onset of the neural process in preparation for an action, subjects can still change their minds and not carry through with the action:

> There could be a conscious "veto" that aborts the performance even of the type of "spontaneous" self-initiated act under study here. This remains possible because reportable conscious intention, even though it appeared distinctly later than onset of RP [readiness potential], did appear a substantial time (about 150 to 200 ms) before the beginning of the movement as signalled by the EMG [electromyogram]. Even in our present experiments, subjects have reported that some recallable conscious urges to act were "aborted" or inhibited before any actual movement occurred.[39]

So, even if Libet's findings were to challenge the existence of "free will," they would not challenge the existence of "free won't," as it is sometimes put.[40] Subjects seem to have control over whether to abort an action they initially intended to perform (or which, as Libet says, they felt an "urge" to perform). And if this is so, then the difference-making cause of the eventual action cannot be the initial neuronal readiness potential alone. It must be—or at least include—some further feature of the subject's mental state.

The observation that subjects can abort an initially intended action is even more relevant when we consider some of the more dramatic follow-up studies inspired by Libet's work. As was described in Chapter 2, John-Dylan Haynes and colleagues used brain-scan data to predict a subject's action several seconds before the action took place. They could do so with an accuracy better than chance but—and this is important—far from perfect.[41] As Eddy Nahmias comments,

> [The] early brain activity predicted a choice with an accuracy only 10 percent better than could be forecast with a coin flip. Brain activity cannot, in general, *settle* our choices [several] seconds before we act,

> because we can react to changes in our situation in less time than that. If we could not, we would all have died in car crashes by now![42]

So, the capacity that is sometimes described as "free won't" seems hard to deny, and this, in turn, implies that the initial neuronal readiness potentials cannot be the difference-making causes of the resulting actions. The subjects' intentional decisions do seem to make a difference.

Of course, one might ask where the ability to abort an initially intended action comes from. Surely one would not wish to invoke a supernatural basis for it, such as a Cartesian "ego" as distinct from any physical properties. So, the ability to abort an intended action must have a physical basis too. In a much-cited study, Marcel Brass and Patrick Haggard identify such a physical basis. They write,

> Our findings suggest that inhibition of intentional action involves cortical areas different from, and upstream from, the intentional generation and execution of action. In addition, this process of "last-minute" inhibition is compatible with a conscious experience of intending to act. Some parts of the inhibition process may occur after the intention to perform an action has become conscious. . . . Libet . . . hypothesized a "veto" allowing the conscious mind to intervene to withhold unconsciously initiated action plans. Because he could not find any identifiable neural correlate of the veto process, he suggested it could involve mind–brain causation: a "free won't" analogous to "free will." Our data identify a clear neural basis for inhibiting intentions and thus identify the neural correlate of the veto process. The hypothesis of a special, non-neural veto process could therefore become unnecessary.[43]

I do not question Brass and Haggard's empirical findings. But I do not think that they contradict the claim that people, as intentional agents, have control over their actions. What their findings tell us is *how* the intentional control mechanism is neurally implemented. That there should be a neural implementation is no surprise. After all, psychological capacities supervene on physical phenomena and do not stand outside the natural order. Crucially, however, this is consistent with the difference-making causes of human actions being found at the level of intentional agency, not at the level of its neural realization. Recall how the control variables of human actions are defined. They are the "parameters which, when changed, lead to systematic changes in other variables of interest," here the actions in question.[44] Recall further that such control variables must "not exhibit gratuitous redundancy," and they must be "manipulable by local processes."[45] Arguably, it is still the subject's (nonaborted) intentions that best fit those criteria, not the underlying neural activity.

Finally, it is worth noting that the Libet-style experiments focus only on relatively simple, spontaneous motor actions, such as hand or finger movements, which are not based on much prior deliberation and certainly do not require any significant planning. Even if those simple motor actions were not always attributable to mental causation, this would not threaten the existence of mental causation in the case of more complex actions. Given what I have argued, it is unlikely that more complex intentional actions will generally admit a satisfactory causal explanation in terms of neuronal readiness potentials alone. Libet recognizes this point:

> In those voluntary actions that are not "spontaneous" and quickly performed, that is, in those in which conscious deliberation (of whether to act or of what alternative choice of action to take) precedes the act, the possibilities for conscious initiation and control would not be excluded by the present evidence.[46]

And Nahmias, similarly, writes,

> The Libet and Haynes research deals with choices that people make without conscious deliberation at the time of action. Everyone performs repetitive or habitual behaviors, sometimes quite sophisticated ones that do not require much thought because the behaviors have been learned. . . . [C]onscious consideration of alternative options . . .

> is a wholly different activity from engaging in practiced routines. A body of psychological research shows that conscious, purposeful processing of our thoughts really does make a difference to what we do.[47]

The bottom line is that the difference-making causes of many human actions remain the agents' intentional mental states. It is at the level of those mental states that systematic "interventions" in matters of deliberative intentional agency are most commonly possible, not at the level of the underlying neural states.

It should be clear, then, where the biggest mistake in the anti-free-will interpretations of the Libet-style experiments lies. It lies in the presumption that, as soon as the psychological mechanisms of intentional action have a neural basis, then there could not be any mental causation. Implicit in this is a kind of causal exclusion argument, which, as I have shown, is flawed. Mental causation is not in tension with the supervenience of "the mental" on "the physical." The claim that the control variables of human actions are people's intentional mental states is perfectly consistent with the assumption that mental states supervene on underlying physical states, to which the ordinary laws of physics apply. In sum, the challenge from epiphenomenalism can be countered. The reality of mental causation is not threatened by the existence of a neural basis for it. Mental states supervene on physical states, but they are not just epiphenomena.

Conclusion

Free will, on the account I have proposed, requires three things: intentional agency, alternative possibilities, and causal control over one's actions. Any organism or entity that meets these requirements has free will. And I have argued that human beings, by and large, meet all three requirements: they are intentional agents, have alternative possibilities, and exercise causal control over their actions. So, they have free will.

In particular, I have defended free will against three broad challenges, which correspond to the three requirements I have introduced. The first challenge—the one from radical materialism—targets the claim that human beings are intentional agents. It asserts that intentional agency has no place in a scientific worldview, suggesting that the notion is a leftover from an outdated folk-psychological way of thinking, to be replaced by a more reductive, neuroscientific account of human behaviour. The second challenge—the one from determinism—targets the claim that human beings have alternative possibilities. It asserts that there could be no alternative possibilities if the world was deterministic; and even if it wasn't deterministic, there could only be randomness, not the kinds of alternative possibilities required for free will. And the third challenge—the one from epiphenomenalism—targets the claim that people have causal control over their actions. It asserts that there is no such thing as mental causation; whenever we act, it is the brain that makes us do it. Although not every free-will sceptic invokes every one of these challenges, they arguably represent a cross-section of the most prominent objections to free will that have been raised from a scientifically informed perspective.

I have shown that all three challenges—while initially formidable—can be rebutted. Although the challenges are quite different from one another, there is one feature they all have in common: Each of them seems to be looking for free will and its prerequisites at the level of the human brain and body, or more generally at the level of the physical system to which the human organism belongs. The challenge in each case lies in the fact that the relevant property—intentional agency, alternative possibilities, or mental causation—is nowhere to be found at that level. From a purely physical or neuroscientific perspective, there are no intentional agents; there are no forks in the road between which an agent can choose; and there is no causation of human actions by people's intentions. There are only physical brains and bodies, as well as law-governed natural processes.

Against this sceptical picture I have argued that free will and its prerequisites are supported by our best theories in the human and social sciences. Intentional agency, alternative possibilities, and causal control are all higher-level phenomena, which emerge from physical processes but cannot be captured in physical terms alone. In this respect, they are in the company of many other familiar higher-level phenomena: beliefs, desires, and intentions; institutions, governments, and cultures; money, inflation, and unemployment. All of these emerge from physical phenomena but cannot be captured in physical terms alone. And, of course, their reality is not in doubt. It would never occur to us to deny the reality of unemployment, or the reality of universities and governments, just because these phenomena are nowhere to be found in fundamental physics. The situation is no different with free will. It is a higher-level phenomenon, but no less real for that.

In closing, I would like to do two things: First, I would like to situate the present picture of free will in the context of the broader philosophical debate on this topic. And second, I would like to ask whether free will is unique to human beings, or whether some other organisms or entities might have it too.

Let me begin with the broader philosophical debate.[1] The most salient distinction in the debate on free will is that between "compatibilist" and "incompatibilist" views. Compatibilism is the view that free will is compatible with determinism, while incompatibilism is the view that it isn't. Let's consider each of these views. Compatibilism comes in different forms. Some compatibilists hold that free will does not require alternative possibilities, but that it requires merely an agent's intentional endorsement of his or her actions. All that matters for free will is that I endorse or "own" my actions; it is not necessary that I could have acted otherwise. Other compatibilists accept that free will requires alternative possibilities but redefine this notion so as to avoid any need for indeterminism. They often do so, for instance, by adopting a conditional or dispositional interpretation of alternative possibilities, as discussed in Chapter 4. Incompatibilists, on the other hand, insist that free will requires alternative possibilities, and hold that this requirement could not be met in a deterministic world. Incompatibilism also comes in different forms. Some incompatibilists think that we truly have free will: the world in which we live provides the right kind of indeterminism. This form of incompatibilism is called "libertarianism." Others think that we have no free will. This might be because the world is deterministic, which would rule out alternative possibilities, or because free will is not compatible with indeterminism either. The view that free will is incompatible both with determinism and with indeterminism is also called "hard incompatibilism."[2] Hard incompatibilists basically think that free will is impossible.

I have called my own view "compatibilist libertarianism," though I have also offered the alternative label "free-will emergentism." At first sight, the claim that a view could be both compatibilist and libertarian sounds paradoxical. After all, libertarianism, as usually defined, is a form of incompatibilism, not compatibilism. As noted, libertarianism is the view that free will requires indeterminism, and that the world is indeterministic in the right way. So, why is "compatibilist libertarianism" not a contradiction in terms?

To show that the label makes sense, let me explain why my view is truly a form of "libertarianism" and yet "compatibilist." Note that, as I have shown, free will requires a form of indeterminism. It requires indeterminism at the agential level: forks in the road between which an agent can choose. Free will is not compatible with agential determinism. Indeed, I have emphasized that some developments in psychology—if they ever occurred—should lead us to give up our belief in free will. Specifically, if new scientific discoveries led us to abandon the picture of human beings as choice-making agents and gave us a deterministic account of human behaviour instead, we would have to conclude that we have no free will. In that sense, the view I have defended is a form of incompatibilism, albeit incompatibilism at the agential level. However, I have also argued that, as of now, the relevant form of indeterminism—namely, agential indeterminism—is supported by our best theories of agency. That is, the human and social sciences, to the best of our understanding, support the claim that there is indeterminism in human agency, which is what matters for free will. In short, the indeterministic requirement for free will is in fact met. For this reason, what I have argued for is not just a form of incompatibilism but also a form of libertarianism. Free will, of the incompatibilist sort I have described, is real.

At the same time, I have argued, all this is compatible with physical-level determinism. Though I have taken no stand on whether the world is deterministic or not at the physical level, I have shown that physical indeterminism is not necessary for free will. That is the sense in which the view I have defended is a form of compatibilism. Putting all of this together yields a form of "compatibilist libertarianism"—one in which libertarian free will emerges as a higher-level phenomenon, which is compatible with lower-level determinism.[3]

My view of free will goes somewhat against the grain in the philosophical debate. While many people outside professional philosophy have incompatibilist intuitions about free will, the majority of professional philosophers are compatibilists. In an influential survey of philosophers, conducted by David Bourget and David Chalmers, 59.1 percent of respondents described themselves as "compatibilists," while only 13.7 percent described themselves as "libertarians." The remaining respondents either thought that there was no free will (12.2 percent) or held other, presumably less standard views (14.9 percent).[4] The view I have defended occupies some kind of middle ground. It vindicates some of the libertarian intuitions that laypeople are often drawn towards while also (hopefully) satisfying the compatibilist leanings of professional philosophers.

Let me now turn to the question of whether free will is uniquely human or whether some other organisms or entities could have it too.[5] This is, of course, a hard question, but I think that the approach taken in this book gives us a way of making it tractable. I have argued that the requirements of intentional agency, alternative possibilities, and causal control encapsulate what it takes to have free will. If this is right, then we should be willing to conclude that *any* organism or entity that meets these requirements has free will. It doesn't matter whether that organism or entity is a human being or something else: whether its "hardware" is human, nonhuman, or not even biological. If you show me an organism or entity that qualifies as an intentional agent, has alternative possibilities, and exercises causal control over its actions, then my analysis suggests that it is a bearer of free will.

Which nonhuman organisms or entities might plausibly be of this kind? The most important examples are certainly nonhuman animals, especially other mammals, and even more so, other primates. As should be clear from my discussion, there is little doubt that many nonhuman animals—especially mammals—qualify as intentional agents, even if their agential capacities are less complex than those of humans.[6] And if we accept that human beings and other animals are similar in other relevant respects, too, then it would seem strange to suggest that alternative possibilities and mental causation are somehow unique to humans. Why should these properties suddenly disappear when we move from the human to the nonhuman case? More likely, the arguments for the reality of alternative possibilities and mental causation apply to nonhuman animals as well—at least to those nonhuman animals that are "usefully and voluminously predictable from the intentional stance," as Dennett puts it.[7] If *Homo sapiens* has free will, then it would seem implausible to suggest that chimpanzees and other great apes don't. And if chimpanzees and other great apes have free will, then it would seem implausible to suggest that other primates don't. Taking this further, insofar as cats and dogs are goal-directed agents which face choices and have conscious states that make a difference to their behaviour, I do not find it preposterous to suggest that they, too, have a certain form of free will, even if it is a much more rudimentary one.[8]

Of course, recognizing that there can be rudimentary forms of free will among nonhuman animals does not entail a denial of important differences between humans and other animals. Without doubt, the agential capacities of human beings are unparalleled in the known biological world. Furthermore, free will should not be equated with responsibility or autonomy in a richer sense. I have suggested that free will is necessary for *fitness to be held responsible*—that is, for being an appropriate candidate for the attribution of responsibility, for praise and blame.[9] Yet, free will by itself is not sufficient. Arguably, a further condition that must be met for fitness to be held responsible is the capacity for normative cognition. Only agents that are capable of normative reasoning about their choices are fit to be held responsible: they must have the capacity to reason about what is "right" or "wrong," "permissible" or "impermissible," "good" or "bad." Only such agents can qualify as *morally responsible* agents, not just as intentional ones. As far as we know, human beings are the only biological creatures that meet this more demanding condition.

What about other, nonbiological entities? It is often claimed, for instance, that certain organized collectives can qualify as intentional agents in their own right, and even as moral agents, over and above their individual members. Philip Pettit and I have defended this view, as have others.[10] Frequently cited examples of group agents are commercial corporations, collegial courts, and other purposive organizations, such as universities or nongovernmental organizations, and on some accounts even states in their entirety. If we look at the conditions for intentional agency, as reviewed in Chapter 3, it should be evident that suitably organized collectives can, in principle, meet them. A firm or other organization can have representational states, which encode its "beliefs" *qua* organization; motivational states, which encode its "desires or goals"; and a capacity to act in a coordinated fashion on the basis of those states. Once we understand an organized collective in this way, we can often make better sense of its behaviour, and predict its actions more reliably, than if we view it simply as an aggregate of many individuals. This is what economists do, for example, when they model firms as rational profit-maximizing actors and explain their market behaviour on this basis. Similarly, scholars of international relations sometimes talk as if entire states, such as the United States or the Soviet Union during the Cuban Missile Crisis, are intentional agents of their own: they have goals that they seek to accomplish (such as increasing their domain of influence); they have beliefs about the means that get them there (be they diplomatic, military, or economic); and they systematically act on the basis of these intentional states (by enacting various policies).[11] In short, some organized collectives may be "usefully and voluminously predictable from the intentional stance," to employ Dennett's criterion one more time.

Some people think that the idea of group agency is just a metaphor. But if we treat certain groups as agents in a literal sense, then we can legitimately ask whether such entities could have free will too. My analysis gives us a way of approaching this question. We must ask whether the relevant collectives, over and above meeting the conditions for intentional agency, also have alternative possibilities and causal control over their actions. Interestingly, the issue of alternative possibilities is seldom discussed in the context of group agents, but it is not absurd to think that there can be agential indeterminism at the level of organized collectives. I do not have the space to develop the argument here, but it is at least conceivable that the arguments for higher-level indeterminism that I have presented (recall Chapter 4) carry over to the case of collectives. And Philip Pettit and I have given an affirmative answer to the question about causal control in group agents.[12] The social analogue of the causal exclusion argument can be rejected for reasons not too dissimilar from those for which we can justifiably reject the exclusion argument against mental causation in the individual case. Arguably, a well-organized group agent does indeed exercise control over its actions. If so, we are faced with the intriguing possibility that group agents—not just their individual members—have free will.

Again, the same qualifications that I made in relation to nonhuman animals apply. Group agents do not match human beings in their agential capacities and sophistication, and they should not be given the same moral status as individual humans. Still, if we think that free will is a necessary condition for fitness to be held responsible, and we wish to argue that group agents are sometimes fit to be held responsible, then we should welcome the conclusion that group agents may have free will. This might reinforce the case for certain forms of corporate civil or criminal liability, something we should surely care about in light of the enormous power that corporations and other collective entities have in the social world.[13] That my account potentially licences the conclusion that there can be "corporate free will" should be seen as a feature of the account, not a bug.

Finally, it is worth mentioning robots and other artificially intelligent systems. There is much to be said about artificial intelligence (AI), and the present book is not the place for a detailed discussion.[14] I simply want to indicate how we might think about free will in this case. First note that many "weak" AI systems engage in relatively specialized number-crunching tasks, and we need not view them as intentional agents. Examples of such systems might be chess-playing computers, Global Positioning System navigation devices, and simple computerized personal assistants that help us perform certain organizational tasks. Even if we can coherently interpret them in intentional terms, this does not seem necessary from an explanatory perspective. We can understand those systems just as well at a purely algorithmic level. And so, it should be clear that they are not candidates for the ascription of free will.

The situation may be different, however, in the case of other, "stronger" AI systems: systems with a richer and more flexible behavioural repertoire and a greater degree of autonomy. Examples might be sophisticated driverless cars, which must flexibly respond to the environment, such as when a child suddenly crosses the road; fully automatic military drones, which select their own targets and make life-and-death decisions autonomously; and, to give a more benign example, future medical helper robots, which take day-to-day responsibility for patients' medical care. Such systems might be best viewed as goal-directed, intentional agents, and not just as number-crunching machines. Viewing them as intentional agents may be not just optional, but explanatorily necessary, in much the same way in which we cannot understand a chimpanzee as a mere biochemical automaton without recognizing its intentional agency. Strong AI systems might therefore pass the test for agency I have proposed (see again Chapter 3) and qualify as intentional agents in their own right, literally and not just metaphorically.

If this is so, then it is reasonable to ask whether those systems might also have alternative possibilities and causal control over their actions. I cannot answer these questions in the abstract, but my account of free will as a higher-level phenomenon offers a way of thinking about them. In particular, it suggests that the implementation of artificial intelligence on a conventional algorithmic computer, with deterministic, mechanical behaviour at a lower level, does not preclude the emergence of intentional agency, alternative possibilities, and causal control at a higher one. The observation that lower-level determinism is compatible with higher-level indeterminism plausibly applies to AI systems as much as it applies to biological organisms. In short, free will need not be tied to any particular hardware. It could occur in social systems and in algorithmic computational ones, not just in biological organisms. Again, I consider it a feature of my account, not a bug, that it tells us something about what it would take to realize free will in systems radically different from a human organism.

Let me conclude with a thought experiment.[15] The kind of compatibilist libertarianism I have defended may seem counterintuitive at first sight. It may seem strange that the world could be deterministic and law-governed at one level, and yet that we could have free will at another. But on further reflection, I think my view is quite attractive. Imagine an article in next week's issue of *Nature* or *Science* that reports a big breakthrough in fundamental physics, especially the discovery that the universe is deterministic. Imagine that, over time, the entire scientific community accepts this finding, and every possible objection is successfully refuted. How should we react to this development?

Should we conclude that it marks the end of free will? Should we think that it challenges our understanding of the human condition as profoundly as the discovery of evolutionary theory did in the nineteenth century, when the traditional picture of life was turned upside down? Should we hope that not too many people find out about it, echoing what the Bishop of Worcester's wife allegedly said when she learnt about Charles Darwin's theory—"Let us hope it is not true, but if it is, let us pray it will not become generally known"?[16] And if the news about determinism became known, would we stop deliberating about what to do and stop holding one another responsible for our actions? Should we perhaps release all murderers from prison, since they could not be held responsible for their actions? Or should we go on with our everyday business, thinking that the new discovery is "just" an interesting development in science?

Surely we ought to do the latter: giving up our conventional understanding of free will and revising the entire fabric of human society would be an overreaction. The approach to free will that I have defended shows why this is so. A relatively mild reorientation of our perspective—namely, a shift in our focus from the physical level to the agential one—is enough to uphold practically everything we conventionally think and say about free will, even against the background of physical determinism and a law-governed universe. If instead we are looking for free will at a physical level, we are looking in the wrong place.

NOTES

REFERENCES

ACKNOWLEDGMENTS

INDEX

Notes

Introduction

1. See Harari (2016).

2. A related argument scheme has been discussed by Nahmias (2014). For a helpful overview of scientific challenges for free will, see also Nahmias (2010).

3. The expression "My brain made me do it" is frequently used in discussions of free will and neuroscience. See, e.g., Bloom (2006), Sternberg (2010), Mackintosh (2011), Szalavitz (2012), and Nahmias, Shepard, and Reuter (2014).

4. I have developed my account of levels more technically in List (2018), building on some earlier articles, including List and Menzies (2009), Menzies and List (2010), List (2014), and List and Pivato (2015). Talk of different levels is common in both science and philosophy. For discussions of levels, see, among many others, Oppenheim and Putnam (1958), Fodor (1974), Owens (1989), Beckermann, Flohr, and Kim (1992), Dupré (1993), Kim (1998, 2002), Schaffer (2003), Floridi (2008, 2011), Ellis, Noble, and O'Connor (2012), Butterfield (2012), and Yoshimi (2012). Levels are sometimes interpreted epistemically (as aspects of our cognition), and other times ontically (as aspects of the world). As I have argued in List (2018), our use of different levels of description renders a levelled ontology plausible.

5. See, for example, Manafu (2015).

6. See, e.g., Bianconi et al. (2013).

7. Technically, we call one level "higher" than another if there is an asymmetrical relation of dependence between the two levels. Specifically, we take the phenomena at one level to be "higher-level" than those at another just in

case the former phenomena "supervene" on the latter and not the other way around. The notion of "supervenience" is explained in what follows. Note that there need not be a linear hierarchy of levels. For instance, while the level of geology and the level of psychology are each "higher" than the fundamental physical level (insofar as both geological and psychological phenomena supervene on physical ones), there is no sense in which geology is "higher" or "lower" than psychology; neither of these two levels asymmetrically depends on the other. For a more formal discussion, see List (2018).

8. On the concept of supervenience, see, e.g., Kim (1984, 1987), Horgan (1993), and McLaughlin and Bennett (2014). In brief, one set of properties (for instance, chemical properties) "supervenes" on another set of properties (for instance, physical properties) if it is impossible for the former to be any different without the latter being different too. Whenever one set of properties supervenes on another, there is a sense in which the former set is dependent on the latter. In the recent philosophical literature, it has become common to analyze relations of metaphysical dependence not (or not primarily) in terms of supervenience but in terms of grounding. See, e.g., Schaffer (2009). For the purposes of this book, however, I prefer to use the metaphysically "lighter" concept of supervenience.

9. See, e.g., Quine (1977) and Fine (1984). I will discuss this in greater detail in Chapter 3.

10. As far as I am aware, my use of the term "compatibilist libertarianism" is new in the literature. I know of only one occurrence of the term in a scholarly publication prior to my own work on free will—namely, in an article on Locke by Rickless (2000); though, given the amount of material that has been written on the topic, it is always possible that I have overlooked some other occurrences. I mention other combinations of compatibilism and libertarianism at the end of this introduction.

11. In an article that came to my attention as I was finalizing this book, Elzein and Pernu (2017) use the term "supervenient libertarianism" to refer to my picture of free will (referring to List 2014, 2015). I have no objection to this term as a possible alternative to "compatibilist libertarianism," although it is unlikely to appeal to those readers who do not like the reference to "libertarianism."

12. I have particularly benefitted from the contributions to the handbook edited by Kane (2002), to which I refer readers for a comprehensive overview of the free-will literature, and I have also benefitted from the debate between

Fischer, Kane, Pereboom, and Vargas (2007). Furthermore, my work is inspired by Van Inwagen's classic contributions (1975, 1983). Although I do not explicitly discuss Van Inwagen's "Consequence Argument" against compatibilism here, I have done so elsewhere (List 2015).

13. See List (2014, 2015), List and Rabinowicz (2014), and List and Menzies (2017). List (2014) was first released in June–July 2011 as an online working paper titled "Free Will, Determinism, and the Possibility to Do Otherwise." A representative lecture that is also available online is "Free Will in a Deterministic World?"; see List (2012).

14. See, e.g., Kenny (1978).

15. See Dennett (2003). Relatedly, see also Dennett (1984) and Taylor and Dennett (2002). Dennett's works have been a great inspiration.

16. See Melden (1961).

17. See Siderits (2008, p. 30). As an aside, I would prefer to speak of "supervenience" rather than "reduction" here. Siderits goes on to write,

> The paleo-compatibilist might use automobile ignition as an analogy. When the key is turned and things go right, we say that the car starts. Suppose someone were to say instead that when the key is turned the contacts in the ignition switch meet, completing the circuit between battery and coil, sending current through the solenoid, thereby inducing current to flow through the starter motor windings. . . . Saying all this has two effects. First, the car has dropped out of the picture. Where we had a car starting, we now have many parts interacting in complex ways. Second, we suspect that perhaps things did not go right. The normal place for talk of the parts and their relations is in diagnosing trouble. We won't find the car and its activity of starting that way. Likewise, the paleo-compatibilist claims, if we want to find moral responsibility we should avoid the temptation to pop the hood. (30)

Intriguingly, Siderits suggests that paleo-compatibilism has its roots in Buddhist thought, and he explores the relevant connections.

18. Relevant works (listed here in the order in which I am about to describe them) include Hoefer (2002), Ismael (2013, 2016), Kane (1998, 2002), Steward (2012), Balaguer (2009), Nahmias (2010, 2014, 2015), Mele (2014, 2017), and Roskies (2006, 2010, 2012). While all of these scholars seek to integrate free will into a scientifically informed worldview, they do so in different ways.

Some of their approaches are best described as "compatibilist," others as "incompatibilist." Hoefer defends a Kant-inspired picture of "freedom from the inside out," according to which the compatibility of free will and determinism can be established if we take an ordinary temporal perspective on the world rather than the atemporal perspective of fundamental physics (in which the focus is on a four-dimensional "block universe"). Ismael argues that "physics makes us free" and understands free will and agency as emergent macroscopic phenomena in a physical world—an idea broadly similar to mine. Kane and Steward each defend incompatibilist positions: Kane argues that indeterminism, at least in relation to certain key actions, is necessary for free will, and suggests that quantum mechanics may support the right form of indeterminism, while Steward argues that agency itself is incompatible with determinism—a claim echoed in some of my discussion of alternative possibilities. Balaguer, too, explores a libertarian picture of free will, and he does so from a naturalistic perspective. For him, a certain kind of indeterminism at the level of the brain is a key prerequisite for libertarian freedom. He treats the question of whether the relevant kind of indeterminism holds as an open scientific question, specifically one for neuroscience and/or physics. Nahmias, Mele, and Roskies, finally, each defend free will against certain neuroscientific challenges. Nahmias argues that, instead of focusing mainly on whether free will is compatible with determinism, the free-will debate should focus more on the challenges raised by neuroscience and psychology, and he suggests that, far from undermining free will, science can help us to explain free will. Mele, similarly, makes a positive case for free will while rebutting some of the neuroscientific challenges, and he does so in a way that remains agnostic on the compatibility of free will and determinism. And Roskies offers a critique of the neuroscientific challenges, arguing that there can be "self-authorship without obscure metaphysics." My analysis of mental causation should resonate with the latter scholars' works.

19. See Vihvelin (2000), Beebee and Mele (2002), and Berofsky (2012). More recently, Arvan (2013) has proposed a distinct position that he also calls "libertarian compatibilism."

20. Examples are the picture of free will as an emergent phenomenon sketched by the physicist Sean Carroll in a blog post, and the views expressed online by Carlo Rovelli, especially the observation that "physical determinism is perfectly consistent with psychic indeterminism"; see Carroll (2011) and

Rovelli (2013). And as I was finalizing this manuscript, Carroll's popular science book, *The Big Picture* (2016), came to my attention, in which, among other things, he defends a view of free will that, like mine, relies on a levelled picture of the world. My own journey to a levelled understanding of free will, especially in my 2011 working paper, predates these publications.

1. Free Will

1. "Free Will," Oxford Dictionaries, https://en.oxforddictionaries.com/definition/free_will.

2. In understanding free will as a three-part capacity, I follow especially Henrik Walter, who, in turn, builds on Gottfried Seebass (1993a, 1993b). Walter writes, "A person has free will (commands freedom of will) if three pivotal conditions are satisfied in a critical number of his acts and decisions. The person: i. *could* have acted *otherwise* (he acts freely), ii. acts for understandable reasons (intelligible form of volition), and iii. is the originator of his actions" (2001, 6).

3. See Kushnir et al. (2015).

4. See Nichols (2006).

5. See Sarkissian et al. (2010).

6. Chernyak et al. (2013, 1343).

7. In line with this, Deery (2015, 2033) has argued that our powerful libertarian free-will intuitions come from our experience of "prospection," defined as "the mental simulation of future possibilities for the purpose of guiding action." He argues that "because of the way in which prospection models choice, it is easy for agents to experience and to believe that their choice is indeterministic." (He suggests, however, that "this belief is not justified," and that the experience of prospection is consistent with determinism.)

8. These passages from Kant's *Groundwork for the Metaphysics of Morals* (G 4:447–8) are quoted in Wood (1999, 175). See also Wood's more general discussion of "freedom as a presupposition of reason" (174 ff.). The idea that when we view someone as an agent we must presuppose that he or she has free will can also be found in Pettit and Smith (1996). They develop this idea by considering the assumptions underlying the "conversational stance" which, they suggest, we adopt when we relate to each other.

9. See Green (2014, 5). There are some kinds of criminal offences which do not fit this simple characterization—for instance, certain statutory offences. I cannot discuss them here.

10. It may be objected that if we hold a person liable for a bad outcome of his or her conduct which he or she failed to notice, then the attribution of liability need not depend on the idea that the person could have acted otherwise; the fact that the failure could be attributed to a bad attitude on the person's part may be sufficient. I suspect, however, that this objection relies on a background assumption of free will. That is, either at the point of action, or at some point in the run-up to the action, or at the point at which the relevant action-supporting attitude was formed, the person could have exercised some free will. It is arguably much harder to sustain the intuition of liability if there was no free will present *anywhere* along the relevant chain of events. I would also like to acknowledge that there are cases of strict liability in tort law. Those are cases of liability where there need not be proof of fault or negligence, such as a manufacturer's liability for certain defective products even when it is not known how the manufacturer caused the defect. It is an interesting question whether free will plays any role in our conceptualization of strict liability.

11. On the relationship between freedom and fitness to be held responsible, see Pettit (2002).

12. See Green (2014, 2–3).

13. See Articles 104 and 105 of the Bürgerliches Gesetzbuch. For discussion, see also Kawohl and Habermeyer (2007).

14. See Baumeister, Masicampo, and DeWall (2009).

15. See Vohs and Schooler (2008).

16. See Shariff et al. (2014).

17. See Sartre (1992, 82), also quoted in Vohs and Schooler (2008, 49).

18. See, respectively, Pereboom (2001) and Harris (2012a). See also Harris (2012b).

19. See Pereboom (2001, 96).

20. For further discussion of responsibility and punishment, see Tadros (2016, especially chapter 5).

21. Smilansky goes so far as to argue that while our beliefs and practices related to free will may rest partly on an illusion, we must retain this illusion. He writes, "The importance of free will for morality in the wide sense and for our view of ourselves is great: it is so great that illusion is required. Illusion

keeps, and ought largely to continue keeping, our moral and personal worlds intact" (2000, 83). Although Smilansky thinks that "there is partial non-illusory grounding for many of our central free will-related beliefs, reactions, and practices," he also suggests that "in various complex ways, we require illusion in order to bring forth and maintain them" (83).

22. For an interesting discussion of free will and its significance, see also Habermas (2007). Although Habermas comes from a somewhat different philosophical tradition, there are a number of resonances between his points and mine. Notably, he takes free will to be "an ineliminable component of our practices of attributing responsibility and holding one another accountable" and "rooted in unavoidable performative presuppositions belonging to agents' participant perspective," which may seem in tension with the "objectivating perspective of the scientific observer" (2007, 13). He suggests that the challenge for a theory of free will is "to reconcile an epistemic dualism of participant and observer perspectives with the assumption of ontological monism" (13), by which he means the assumption that "the universe is unified and includes us as part of nature" (14).

23. In the Freedom House rating, North Korea has received a score of 7 ("least free"). See Freedom House (2015).

24. For surveys and further discussion, see, e.g., Kramer (2003), Carter (2012), and List and Valentini (2016).

25. Or, to take another division in the debate on social freedom, some people think that constraints on action that are sufficiently justified—perhaps constraints in the name of social equality—pose no threat to social freedom, while others think that *any* constraints on action restrict freedom, whether justified or not. On the former view, the fact that the state coerces me to pay taxes does not restrict my freedom provided the system of taxation is just. On the latter view, it does: any instance of state coercion reduces my freedom, however justified it may be. To give a further example, some people think that a person is socially free provided he or she is *contingently* able to lead a relatively unconstrained life with decent opportunities. Others think that social freedom requires more than this. Consider, for instance, a benign dictatorship—say, a free-market technocracy with a high standard of living but no democratic rights. Such a regime might give its citizens a fair range of opportunities and yet have the power to withdraw them at its discretion, thereby making those opportunities contingent on

the power and goodwill of the regime. The observation that such opportunities are insufficiently protected leads some to think that real freedom requires more than opportunities and noninterference; it requires "nondomination": the *robust* or *guaranteed* absence of relevant impediments, as Pettit (1997, 2002) defines it.

26. Of course, the issues here are complicated, and—depending on how exactly we define social freedom—not every increase in social freedom will necessarily give people more room to exercise their capacity of free will in valuable ways. An increase in pure market freedom, for instance, may turn out to compromise individual freedom in some other respects. These issues go beyond the scope of this book.

27. In relation to this three-part structure, recall my references to Walter (2001) and Seebass (1993a, 1993b).

28. See Edwards (1754, 28).

29. The philosopher Frankfurt (1969) has constructed some ingenious examples to challenge the idea that the possibility of acting otherwise is necessary for moral responsibility. My own view is that these examples ultimately do not succeed. Here, in brief, is my objection. Consider the following generic Frankfurt-style example: Jones is deciding whether to kill Smith. Unbeknownst to Jones, an evil manipulator has implanted a device into Jones's brain through which the manipulator can control Jones. Specifically, the device is set up to monitor Jones's decision process. If Jones independently comes to the decision to kill Smith, then the device does nothing; everything in Jones's brain continues to work as if the device were absent. However, if Jones leans against killing Smith, then the device intervenes and makes him "decide" to kill Smith. Now, suppose that it so happens that Jones makes the killing decision by himself, without any intervention from the manipulative device, and he goes ahead and kills Smith. Frankfurt notes that we have the strong intuition that Jones *is* responsible for the killing. After all, it seems that he clearly qualifies as the "author" or "maker" of his decision. And yet, it also seems, he did not have alternative possibilities. He could not have acted otherwise. Acting otherwise would have been prevented by the manipulative device. I agree with the intuition, but I do not think that it shows that responsibility does not require some form of indeterminism. For the example to be compelling, it has to be the case that there are two alternative possibilities: either Jones makes the killing decision by himself, or he does so through the device's intervention; call the first case the "voluntary case" and the second the "interven-

tion case." Admittedly, those are not alternative possibilities of *action,* but they are still alternative possibilities in relation to a certain *mental act.* Even though Jones cannot act otherwise than he eventually does (by refraining from the killing), there is a kind of fork in the road for Jones. One prong of that fork corresponds to the voluntary case, the other to the intervention case. I contend that unless both prongs are genuine possibilities for Jones, the intuition that he should be held responsible is significantly weakened. For a more elaborate defence of the thesis that alternative possibilities are necessary for moral responsibility, see Alvarez (2009). Alvarez, too, questions the "conceptual cogency" of Frankfurt's examples. She argues that those examples "require a counterfactual mechanism that *could* cause an agent to perform an action that he cannot avoid performing" and that "given our concept of what it is for someone to act, this requirement is inconsistent" (2009, 61).

30. See Dennett (1984). A nice discussion of this story can also be found in Kane's introductory article in Kane (2002).

31. I have previously discussed this interpretation of the Luther story in List (2014) and List and Rabinowicz (2014).

32. In a similar vein, Alvarez (2013) describes the "the idea of a causal power, a power to cause things" as central to human agency and argues that when we act, we must have a "'two-way' causal power" over our action (2013, 103 and 101).

33. Or, at least, there must be some description of the action under which we intended to do it. Knobe (2003) gives the example of a company manager who decides to implement a new industrial project so as to make a lot of money for his company. As a side effect, of which the manager is fully aware, the project will harm the environment, but the manager cares only about the money and not about the environmental harm. Even though, strictly speaking, the manager only intends to make money and doesn't intend to harm the environment, we may still want to attribute the act of harming the environment to the manager's free will. Arguably, it is true that the manager intended to perform that act *under some description*—namely, the description that focuses on the money.

34. See Davidson (1980, 79).

35. Along similar lines, Walter writes,

> These three features [in his own formulation, quoted above] rest on central intuitions that can be found in nearly all theories about free

> will. Those various theories differ solely in the fact that they either deal only with part of the components, or they declare one of them to be particularly significant, or they support variously strong interpretations. (2001, 6)

A subtly different, though related, account of three dimensions of free will can be found in a paper by O'Leary-Hawthorne and Pettit (1996); they distinguish between "freedom as ownership," "freedom as underdetermination," and "freedom as responsibility."

36. See England (n.d.).

37. For a detailed analysis of what it means to be the agent of some action, see Himmelreich (2015).

2. Three Challenges

1. See, for example, P. M. Churchland (1981) and P. S. Churchland (1986).

2. See Heider and Simmel (1944).

3. See Pyysiäinen (2009). The sceptic Shermer calls this tendency "agenticity": "the tendency to believe that the world is controlled by invisible intentional agents" (2009).

4. It is important to note that, although some neuroscientists may hold views along these lines, research in neuroscience need not be committed to this radical thesis. A more moderate view is that neuroscience complements, rather than replaces, more traditional psychology, and thereby sheds light on the neuroscientific foundations of intentional agency.

5. See P. M. Churchland (1981, 75).

6. See P. M. Churchland (1981, 73).

7. See P. M. Churchland (1981, 85).

8. For instance, in a recent neuroscientific article on action control and the brain, Uithol, Burnston, and Haselager argue that "we may not find intentions in the brain":

> Intentions are commonly conceived of as discrete mental states that are the direct cause of actions. . . . [T]he processes underlying action initiation and control are considerably more dynamic and context sensitive than the concept of intention can allow for. Therefore, adopting the notion of 'intention' in neuroscientific explanations can easily lead

to misinterpretation of the data, and can negatively influence investigation into the neural correlates of intentional action. (2014, 129)

9. See Steward (2012, 10). In her description of the challenge, Steward further notes that "the huge success of molecular biology is another important factor, providing evidence . . . that at least some of the complex, higher-level phenomena associated with life in all its manifestations are susceptible to reductive explanation at the lower level represented by chemistry" (10).

10. See Harmon (2010), reporting on Buckholtz et al. (2010).

11. See, e.g., Settle et al. (2010), Hatemi et al. (2011), and Ebstein et al. (2015).

12. See DeWitt, Aslan, and Filbey (2014, 157).

13. See, e.g., Greene et al. (2001).

14. See Kahneman (2011).

15. For a more detailed discussion of these points, on which I have here drawn, see Ramsey (2013, especially section 3.1).

16. For a classic account of this challenge, see Van Inwagen (1975, 1983).

17. I have previously discussed this argument in List (2014). Vihvelin (2013, 2) calls this the "Basic Argument" for incompatibilism about free will.

18. Strictly speaking, what I have defined here is the *initial segment* of the world's history up to a particular point in time. A *history in its entirety* is the sequence of states of the universe across *all* points in time—past, present, and future.

19. I will discuss all of these notions in more detail in Chapter 4. For formally precise versions of the present definitions, see List (2014) and List and Pivato (2015). For more on the definition of determinism, see also Müller and Placek (2018).

20. For this frequently quoted 1820 passage from Laplace, see, e.g., L. Smith (2007, 3).

21. On the relationship between contemporary physics and free will, see, e.g., Bishop (2002) and Hodgson (2002).

22. The name comes from the workplace of some of quantum theory's pioneers.

23. For an accessible discussion, see Musser (2015).

24. See, e.g., Kane (1998). The physicist Penrose (1994) argues that the brain might be, in some respects, a quantum computer, suggesting that it makes use of quantum indeterminacies through so-called microtubules. It is fair to say, however, that Penrose's ideas remain controversial. For arguments

to the effect that quantum indeterminacies are irrelevant for free will, see, e.g., Honderich (2002) and Dennett (2003).

25. See Hoefer (2002, 202).

26. See, e.g., Kim (1998, 2005).

27. See Harris (2012a, 7–8).

28. This quotation comes from an interview conducted by Cook (2011). See also Gazzaniga (2011).

29. See Libet et al. (1983).

30. See Wegner (2002, 55).

31. See Haynes et al. (2007).

3. In Defence of Intentional Agency

1. My discussion of agency draws on List and Pettit (2011, chapter 1). For a seminal account of intentional agency, see Bratman (1987). The notion of a "belief-desire-intention agent," common in computer science, builds on Bratman's account. The basics of the present picture of agency go back to David Hume.

2. In my own decision-theoretic work, together with Franz Dietrich, I have emphasized the idea that an agent's choice behaviour is to be explained not merely in terms of the agent's preferences and beliefs, but in terms of some additional features of the agent's psychology. Our focus has been on motivating reasons, which may in turn depend on the context the agent finds him- or herself in, but in reality human psychology is obviously even richer than that. See, e.g., Dietrich and List (2016b).

3. See, e.g., Searle (1979, 1983), Dennett (1987), Dennett and Haugeland (1987), and Siewert (2017).

4. I am here glossing over some further complications. Arguably, there are two respects in which intentional states can be *about* something. First, they have *contents*. If I believe that there is coffee available in the kitchen, then the content of that belief is the proposition that there is coffee available in the kitchen. But intentional states can also have *objects*. The *objects* of my belief that there is coffee available in the kitchen are the coffee and the kitchen. See also Searle (1979).

5. See Dennett (1987, 17).

6. See Dennett (2009, 339).

7. See, e.g., Dennett (2009).

8. See Dennett (2009, 342). The first semicolon in the quoted passage replaces a comma in Dennett's original text.

9. For a similar point—in relation to the attribution of cognitive representations to a system—see Van Gelder:

> A useful criterion of representation—a reliable way of telling whether a system contains them or not—is to ask whether there is any explanatory utility in describing the system in representational terms. If you really can make substantially more sense of how a system works by concretely describing various identifiable parts or aspects of it as representations in the above sense, that is the best evidence you could have that the system really does contain representations. Conversely, if describing the system as representational lets you explain nothing over and above what you could explain before, why on earth suppose it to be so? (1995, 352)

Relatedly, see also Dretske (1999).

10. My disagreement with interpretivism is similar to that of Steward, who writes, "We have to treat certain things as agents, roughly speaking, because that is what they are, and not the other way around" (2012, 106). For Steward, interpretivism of the kind associated with Dennett "misidentifies the correct direction of explanation" (106).

11. For a detailed survey of the philosophical debate about the cognitive capacities of nonhuman animals, see Andrews (2016). I can obviously not do justice to all the nuances of this debate here.

12. For a recent study of the brain size of dogs, from which the present estimate is taken, see Jardim-Messeder et al. (2017). Obviously, all such estimates have to be interpreted with a grain of caution.

13. This is the "binding problem," on which there has been a fair amount of neuroscientific research in recent decades. See, e.g., Treisman (1996) and Singer (2007).

14. My points here echo those famously made by Dennett (1987, 2009). I have also made some similar points in my joint work with Dietrich in defence of mental-state realism in economics (Dietrich and List 2016a, section 8). Technically, as anticipated in my defence of the simple test for intentional agency, the present argument for ascribing intentional agency to

nonhuman animals is an inference to the best explanation. As Andrews (2016) puts it,

> The inference to the best explanation argument justifies the attribution of mental states to animals based on the robust predictive and explanatory power that is gained from such attributions. As the argument goes, without such attributions we would be unable to make sense of animal behaviour. This argument relies on ordinary scientific reasoning; of two hypotheses, the one that better accounts for the phenomenon is the one to be preferred. Those who offer this sort of argument for animal minds are claiming that having a mind (whatever that amounts to) better explains the observed behaviour.

In a similar vein, Steward defends the attribution of agency to many animals, while noting that it is ultimately an

> empirical question . . . whether the [nonagential] algorithmic type of approach to what looked at first sight to be flexible behaviours in [a creature] might succeed. If it did . . . , then . . . we would have no reason to continue to maintain the view of that creature as an agent. . . . If it did not, then the folk-psychological interpretation would have been saved. But what counts as "success"? . . . Would the various possible mechanistic approaches to [the creature's] movements have to become too complex, too overburdened by numbers of variables, too bogged down with exception clauses, in order to accommodate all the data concerning its behaviour? (2012, 112–113)

In light of these questions, it seems that the case for the hypothesis that a given creature is an agent must involve some kind of inference to the best explanation.

15. For overviews, see, e.g., Camerer, Loewenstein, and Rabin (2004) and Kahneman (2011).

16. This is the "structure-agency debate" in the social sciences. For helpful overviews, see, e.g., Sewell (1992) and Haslanger (2016).

17. As Dennett and Haugeland noted more than thirty years ago, "Dispensing with intentional theories is not an attractive option . . . , for the abstemious behaviorisms and physiological theories so far proposed have

signally failed to exhibit the predictive and explanatory power needed to account for the intelligent activities of human beings and other animals" (1987, 385). This observation is no less true today.

18. See, e.g., Treisman (1996) and Singer (2007).

19. For a thought-provoking overview, see Tononi and Koch (2015).

20. See Rutledge et al. (2015). For an earlier overview, see also Zuckerman and Kuhlman (2000).

21. There are some possible complications here which I have set aside for simplicity. Even in cases where we would conventionally say that a higher-level property $\mathbb{P}$ can be successfully reduced to a corresponding lower-level property *P*, as in the example of temperature and mean kinetic energy, the lower-level property will typically be an adequate substitute for the higher-level one only for certain carefully demarcated explanatory purposes. So, strictly speaking, when we say that *P* "can serve as a substitute" for $\mathbb{P}$, we must specify for which explanatory purposes this is so. Substitutability will often be restricted to what are called "extensional contexts." It is widely recognized, for instance, that substitutability may fail in "nonextensional contexts," such as those involving mental state ascriptions. Even if, in our example of temperature and mean kinetic energy, property *P* is an excellent substitute for property $\mathbb{P}$ in causal explanations, substitutability may still fail when we are engaged in discourse about the beliefs that someone has about these properties. Suppose we are trying to explain Mr. X's beliefs about the temperature, and we have figured out that Mr. X believes that property $\mathbb{P}$ holds ("the temperature is such-and-such"). It will not necessarily be true that Mr. X also believes that property *P* holds ("the mean kinetic energy is such-and-such"), because Mr. X may be unaware of the equivalence between temperature and mean kinetic energy.

22. The quotation comes from Siewert (2017).

23. See, e.g., Searle (1983).

24. For discussion, see Bennett et al. (2007).

25. My discussion of this example closely follows List and Spiekermann (2013, especially 634).

26. See, e.g., Fodor (1974) and Putnam (1975b).

27. For the record, Polger and Shapiro (2016) have recently criticized the idea that mental states are multiply realizable. For a discussion of the strengths and weaknesses of their critique, see Levin (2016).

28. For a relevant discussion, see Quian Quiroga, Fried, and Koch (2013).

29. I am borrowing the example of the relationship between the pixels and the image from Lewis (1986, 14). Pettit and I have also used the example in our work on group agency (List and Pettit 2011, 65).

30. Formally, one set of properties (the "higher-level" properties) is said to "supervene" on another (the "lower-level" properties) if fixing the second set of properties necessarily fixes the first. In our example, the locations and colours of the pixels are the lower-level properties. The features of the image, such as the depicted hairstyle and facial expression, are the higher-level properties.

31. The present wording and some of the subsequent discussion are based on List (2014, 167). The naturalistic ontological attitude goes back at least to Quine's naturalistic realism (e.g., 1977). For a detailed discussion, see Fine (1984).

32. I have made a related case for mental state realism in my joint work with Dietrich (Dietrich and List 2016a).

33. Note that we cannot observe electromagnetism and gravity themselves; we can only observe their manifestations in the behaviour of other objects. For a defence of a more sophisticated successor to instrumentalism, called "constructive empiricism," see Van Fraassen (1980).

34. See Putnam (1975a, 73), cited in Chakravartty (2011).

4. In Defence of Alternative Possibilities

1. Specifically, the traditional Copenhagen interpretation takes quantum mechanics to entail some form of physical indeterminism. It is important to note, however, that the tension between quantum mechanics and general relativity arises independently of whether we adopt this interpretation of quantum mechanics.

2. In what follows, I draw on the analysis in List (2014).

3. See Moore (1912), and subsequently Ayer (1954), among others.

4. See Moore (1912, 12).

5. See, for example, M. Smith (2003), Vihvelin (2004), and Fara (2008), as well as critical discussions in Clarke (2009), Whittle (2010), and Berofsky (2011).

6. Generally, someone may be severely constrained in what he or she can do under the conditions that actually obtain, and thus lack alternative pos-

sibilities in commonsense terms, and yet have an unexercised disposition to act otherwise in some other circumstances. See also Whittle (2010).

7. See Whittle (2010, 21).

8. To be sure, those two interpretations are not completely far-fetched: they capture something that matters for agency. When we assess someone's agential capacities, we may well be interested in what would happen if he or she were to try to do otherwise or in how the person would be disposed to act under various conditions. Both questions are relevant to the study of agency, and I will return to them when I discuss whether agents have causal control over their actions. But neither the conditional nor the dispositional interpretation captures the sense of alternative possibilities required for free will and responsibility.

9. See Hurley (1999, 205–206, emphasis in the original).

10. For an accessible discussion of Einstein's comment, see Musser (2015). Note that the definition of determinism may require some fine-tuning if the distinction between future and past is not absolute, but relative to a reference frame, as in Einstein's theories of relativity. I set these complications aside here.

11. For comparison, take the standard argument: (1) all humans are mortal, (2) Socrates is human, therefore (3) Socrates is mortal. This is valid. Were we to replace the first premise (1) with the premise that all mammals are mortal, the argument would no longer be valid, unless we add the further premise that all humans are mammals.

12. As mentioned in the Introduction, my claim that we can avoid the incompatibilist conclusion by keeping different levels separate echoes earlier points made by Melden (1961), Kenny (1978), and Siderits (2008).

13. I first presented this argument in List (2014); a related idea can be found in Dennett (2003). Dennett argues that determinism (which I would understand as physical determinism) does not imply inevitability (which I would understand as agential determinism). However, he does not make the notion of "evitability" fully precise. In a joint discussion, Taylor and Dennett (2002) go some way in that direction, suggesting that, in statements such as "it is possible for an agent to do otherwise," we should use a "broad" (permissive) method for identifying the relevant possible worlds, not a "narrow" (restrictive) method. This enables us to say that it is possible for an agent to do otherwise in a deterministic world. But this analysis is arguably more ad hoc and less systematic than the one I am about to present. Similarly, as noted,

Kenny (1978) argues that determinism at a lower, physiological level is compatible with free will at a higher, psychological one. But he does not develop these ideas in formally precise terms and vacillates a bit between different interpretations of alternative possibilities (dispositional and conditional ones, as explained in List 2014, 175–176n15).

14. This is not to deny that, if we were to switch to the physical level, there would be a fact about the physical-level state. At that physical level, however, we would no longer be talking about agency.

15. For a more technical version of the model, see List (2014). The idea of branching histories, which is central to my analysis, has also been employed by other authors in the context of free will. See, e.g., Belnap (2005).

16. In the case of a stochastic system, the laws may further need to specify probabilities. I set these technicalities aside here. For a formal analysis of dynamic and stochastic systems, see List and Pivato (2016).

17. See Butterfield (2012, 108, emphasis removed). See also List and Pivato (2015) and, for formally related results, Werndl (2009). Werndl raises the intriguing possibility that deterministic and indeterministic descriptions of a system might generally be observationally equivalent. A system that appears deterministic at one level of description might have indeterministic underpinnings at another level, and vice versa. The point that a system may behave deterministically at one level, while at another level displaying dynamics that can be interpreted probabilistically is also familiar from discussions of whether there can be objective probability in a deterministic world. See, among others, Von Plato (1982), Loewer (2001), Ismael (2009), Glynn (2010), Sober (2010), Strevens (2011), and Frigg and Hoefer (2015). For an analysis of the relationship between lower-level and higher-level phenomena using dynamical systems theory and a critical discussion of the relationship between physical and mental phenomena in this context, see Yoshimi (2012).

18. My claim that psychology is indeterministic—and that this is consistent with physical determinism—may remind some readers of Davidson's famous thesis of the "Anomalism of the Mental," in which he argued that "there are no strict deterministic laws on the basis of which mental events can be predicted and explained" (2001, 208, originally published in 1970), despite the supervenience of the mental on the physical. Here, however, I do not rely on Davidson's arguments but offer my own arguments for indeterminism in psychology. A comparison of my views with Davidson's is beyond the scope

of this book. For discussions of Davidson's "anomalous monism," see, e.g., Yoshimi (2012) and Yalowitz (2014).

19. The wording in this paragraph draws on List (2014, 171).

20. For a related discussion, see Maier (2015). Maier develops an account of the "agentive modalities," in which the decision-theoretic notion of an "option" is the key primitive notion, and he also explores implications for the free-will debate.

21. This is so unless we are willing to redescribe the options in an artificially gerrymandered way. For discussion, see Dietrich and List (2016b).

22. In Kahneman and Tversky's experiments, subjects were presented with different choice situations in which the same menu of options was given to them under subtly different descriptions. It turned out that changes in the description of the options were sufficient to change people's choice behaviour, even though the options remained objectively the same. We would not be able to explain such patterns of choices if the participants each simply had a fixed preference ordering over the options. These experiments, in turn, led to the insights underlying the influential idea of nudging, now popular among some policy makers. For an overview, see, e.g., Kahneman (2011).

23. Steward (2012) defends a view she calls "agency incompatibilism," according to which there can be no agency in a deterministic world. My argument here can be read as supporting a kind of agency incompatibilism, where the incompatibility, crucially, is between agency (at least of a nontrivial sort) and agential (rather than physical) determinism.

24. The wording of my example of decision and game theory draws on List (2014, 168). In a working paper on free will, the economist Gilboa (2007) acknowledges that rational choice theory implicitly assumes free will but raises questions about whether this can be reconciled with a scientific worldview. Here I aim to achieve such a reconciliation.

25. The wording in the last two paragraphs draws on List (2014, 169).

26. The present section and the next one are based on List (2014, sections 7 and 8).

27. I am grateful to Ann Whittle for helpful comments prompting me to explore the relationship between my analysis and Kratzer's semantics of "can." See Kratzer (1977). On the semantics of "can," "options," and "abilities," see also Maier (2015).

28. Slightly more precisely put, the proposition expressed by the embedded nonmodal sentence "*A* does *B*" must be consistent with those constraints.

29. See Kratzer (1977, 342–343).

30. In fact, this challenge is widely recognized in the debate on libertarian conceptions of free will in, among others, Strawson (1994), Kane (1999), and Mele (2017, chapter 10), as briefly discussed later in this section. I have discussed the challenge and the response to it in joint work with Rabinowicz, on which I here draw; see List and Rabinowicz (2014).

31. See Kane (1999, 217).

32. See, respectively, Strawson (1994), Mele (2017, with the quotation taken from 206, emphasis added), and Kane (1999, with the quotation taken from 224; the notion of a "self-forming act" was introduced by Kane). For a discussion of the relative merits of Mele's and Kane's responses to the challenge, see also Mele (2017, chapter 10). Note that Mele develops this response to the challenge without actually endorsing a libertarian view of free will. Needless to say, Strawson's, Mele's, and Kane's arguments are more sophisticated than what my brief summaries may suggest.

33. See List and Rabinowicz (2014, 155).

34. One might raise a further objection. Suppose we grant that, despite agential indeterminism, the street robber *is* responsible for his action because—or at least partly because—he intentionally endorses it. Should we not worry about the source of the intentional endorsement here? Why does an agent intentionally endorse his or her action? The answer is presumably that the action is supported by the agent's goals and plans at the time. But where do these goals and plans come from? Is the agent responsible for them as well? Perhaps there was an earlier choice that led to the formation of those goals and plans, where the outcome of that earlier choice was itself one the agent intentionally endorsed. However, this line of reasoning quickly runs into trouble. If we look at the relevant antecedents of any given action—even one that results from the most careful process of rational deliberation by the agent—then sooner or later we are likely to find some antecedents of the action that do not plausibly fall under the umbrella of the agent's responsibility. The street robber, for instance, may well intentionally endorse his action now—indeed, the action may be fully rationalized by his beliefs and desires at this time—but is he also responsible for those beliefs and desires themselves, the mind-set without which he would never have chosen to commit

the robbery? Suppose we accepted the principle that an agent cannot be responsible for an action unless he or she is also responsible for the underlying action-supporting intentions, where responsibility for those intentions would, in turn, require that these be attributable to a responsible choice too, one which satisfied similarly demanding conditions. Then we might have to conclude that people are responsible for far fewer actions than commonly assumed, and perhaps there is very little scope for responsibility at all. My own view is that the principle leading to this conclusion is simply too demanding. The street robber, for instance, *can* be responsible for his action, even if he is not responsible for all of the prior steps within the agential history that put him on track to becoming a street robber. Of course, some aspects of his personal history may indeed reduce his overall culpability, and a judge may legitimately take evidence of that history into account in adjudicating the case. Personal hardship, a difficult social environment, and other factors may well be considered to reduce culpability. That said, if the street robber committed his action intentionally, had alternative possibilities, and was in control of what he did, then there is no reason not to treat this as a case of free will in the sense discussed in this book, even if we reach a more nuanced assessment of the robber's culpability in light of his prior history. If one wished to develop a more demanding account of responsibility under which an agent's responsibility for his or her action-supporting intentions is central, then one might try to import a suitably adjusted variant of Robert Kane's idea of "self-forming actions" into my compatibilist libertarian framework (see, e.g., Kane 1999). Exploring this theoretical possibility, however, is beyond the scope of the present discussion.

5. In Defence of Causal Control

1. This must be so at least under some description of the action. Recall my earlier reference to Knobe (2003).

2. See Harris (2012a, 7).

3. See Hume ([1739–40] 1888, book 1, part 3, section 6, 93).

4. In distinguishing these three interpretations, I am following Garrett's (2009) discussion of Hume.

5. See Kim (1998, 2005).

6. I here build on my joint work with Menzies. See, in particular, List and Menzies (2017).

7. By "physical effect" I mean a physical event that has some cause. Perhaps *every* physical event is a physical effect in this sense, in that there are no physical events that are uncaused. For present purposes, however, I can leave this possibility open and focus only on those physical events that have *some* cause. The causal closure principle as stated here is less demanding than the stronger principle that says that every physical event has some physical cause. Even though many proponents of a scientific worldview might be ready to endorse that stronger principle too, it is not needed for the causal exclusion argument that I am discussing.

8. As explained below, we may amend this principle by recognizing that there may be exceptions to this rule—namely, cases of *genuine causal overdetermination.*

9. Two things are worth emphasizing. First, we are not considering *merely partial* causes here, of which several might well be needed to bring about the effect in question; we are considering *wholly sufficient* causes. Second, a given effect could have distinct causes *at different times,* where earlier causes were sufficient to bring about the later ones, as in a sequence of events in which one thing causes another. What the exclusion principle captures is the idea that it is generally unparsimonious to attribute a single effect *simultaneously* to two distinct causes, where each is *the wholly sufficient cause* of the effect.

10. Such cases have received a fair amount of attention in the philosophical literature, and they raise interesting questions for the notion of causation, but I will set them aside here. For a helpful discussion, see, e.g., Hall (2004).

11. The present case, in which an intention and the underlying physical state compete for causal relevance, is not plausibly a case of genuine causal overdetermination, as introduced earlier. In genuine overdetermination cases, the competing causes are either independent from one another or, at most, contingently connected (such as the simultaneous shots fired at some human target by two or more assassins). The agent's intention, unlike in those overdetermination cases, presumably *supervenes* on the underlying physical state and is therefore necessarily (not just contingently) connected to it.

12. On connections between the problems of mental causation and free will, see also Bernstein and Wilson (2016).

13. My discussion here follows List (2018, section 4.1).

14. My discussion of what I have called problem 1 follows List (2018, section 4.1). On the question of whether there is a fundamental level, see also Schaffer (2003).

15. See Block (2003, 138).

16. See Kim (1993, 337), also quoted in Schaffer (2003).

17. For a critical discussion, see Schaffer (2003).

18. See Block (2003, 138).

19. See Russell (1913, 1).

20. See Russell (1913, 14).

21. See Woodward (2009, 257).

22. In a similar vein, Norton describes the notions of cause and effect as "heuristically useful," but warns us not to "mistake them for the fundamental principles of nature" (2003, 2). Norton defends the idea of "causation as folk science." Importantly, he does not think that causal reasoning is misguided altogether. He summarizes his argument as follows:

> I will characterize causal notions as belonging to a kind of folk science, a crude and poorly grounded imitation of more developed sciences. More precisely, there are many folk sciences of causation corresponding to different views of causation over time and across the present discipline. While these folk sciences are something less than our best science, I by no means intend to portray them as pure fiction. Rather I will seek to establish how their content can be licensed by our best science, without the causal notions' becoming fundamental. (2)

This picture has some commonalities with the picture of causation as a special-science phenomenon that I take to be the correct one.

23. See Cartwright (1979) and Field (2003, 435), respectively. For a pertinent discussion of the indispensability of causal reasoning in the special sciences and a helpful historical summary, see also Ismael (2016, chapter 5), as well as Ismael (2013).

24. See, for example, Lewis (1973), Menzies and Price (1993), Pearl (2000), Woodward (2003), and Halpern and Pearl (2005a, 2005b). For an overview, see Menzies (2014).

25. See List and Menzies (2009). See also List and Menzies (2017).

26. See, e.g., Woodward (2008), Raatikainen (2010), Campbell (2010), Roskies (2012), and Ismael (2013, and 2016, chapter 5). Relatedly, see Deery

and Nahmias (2017). In fact, my arguments in this section are in line with what is increasingly becoming the mainstream view among proponents of an interventionist theory of causation.

27. See Jackson and Pettit (1990).

28. On the notion of realization insensitivity and its significance, see List and Menzies (2009).

29. See Roskies (2012, 329). The reference is to Campbell (2010).

30. See Roskies (2012, 325).

31. See Campbell (2010, 26).

32. For details, see List and Menzies (2009).

33. I should note that the causal exclusion argument in its original form arguably relied on an understanding of causation distinct from the two disambiguations of "sufficient cause" that I have considered—namely, an understanding of causation as "production." On such an understanding, to be the cause of some effect is to be the "producer" of that effect, in a suitable metaphysical sense of production. The metaphor that is usually invoked to convey this idea is the metaphor of the "causal oomph." A cause produces its effect via some causal force or power—its "oomph"—just as a billiard ball causes the movement of another ball through a forceful impact. Unfortunately, however, the production account of causation remains relatively vague and is out of line both with the practice of causal interference in science and with the recent technical literature on causal modelling (which, as noted, focuses on difference making). For a comparison of production and difference-making accounts of causation, see Hall (2004).

34. See Libet et al. (1983).

35. See Haggard, Clark, and Kalogeras (2002, 382).

36. See, e.g., the helpful discussions in Roskies (2010), Mele (2014), and Nahmias (2015).

37. See Mele (2014, 24–25).

38. Mele (2014, 12).

39. See Libet et al. (1983, 641).

40. For discussion, see also Shermer (2012).

41. See Haynes et al. (2007).

42. See Nahmias (2015, 78).

43. See Brass and Haggard (2007, 9144).

44. See Roskies (2012, 329), drawing on Campbell (2010).

45. See Roskies (2012, 325).

46. See Libet et al. (1983, 641).

47. See Nahmias (2015, 78).

Conclusion

1. As already noted, an overview can be found in the handbook edited by Kane (2002). For another survey, see O'Connor and Franklin (2018).

2. See, e.g., Pereboom (2001).

3. As noted earlier, other works that combine compatibilist and libertarian ideas, albeit in different ways and under different labels (such as "libertarian compatibilism" or "Humean compatibilism"), include Vihvelin (2000), Beebee and Mele (2002), Berofsky (2012), and Arvan (2013).

4. See Bourget and Chalmers (2014).

5. For a discussion of agency in nonhuman systems (including machines, animals, and even plants), see Dretske (1999).

6. On animal agency, see Steward (2009).

7. And, we might add, where taking the intentional stance is explanatorily necessary.

8. On free will in animals, see Steward (2015). My suggestion that some nonhuman animals can not only qualify as agents but also have alternative possibilities and mental causation is broadly consistent with Steward's suggestion that some animals can have "two-way powers."

9. My understanding of the notion of "fitness to be held responsible" draws on Pettit (2002, 2007) and List and Pettit (2011, chapter 7).

10. See, e.g., French (1984), List and Pettit (2011), and Tollefsen (2015).

11. See, e.g., Waltz (1979) and Snidal (1985).

12. See List and Pettit (2011, chapter 7).

13. On corporate responsibility, see also French (1984), Erskine (2001), and Copp (2006), as well as Pettit (2007) and List and Pettit (2011, chapter 7).

14. Helpful overviews can be found in Russell and Norvig (2009), Winfield (2012), and Boden (2016).

15. The concluding thought experiment is taken from List (2014), with slightly adapted wording.

16. This is taken from Jerry Coyne's opening quotation in the Introduction.

References

Alvarez, M. (2009). "Actions, Thought-Experiments and the 'Principle of Alternate Possibilities.'" *Australasian Journal of Philosophy* 87(1): 61–81.

Alvarez, M. (2013). "Agency and Two-Way Powers." *Proceedings of the Aristotelian Society* 113(1, pt. 1): 101–121.

Andrews, K. (2016). "Animal Cognition." In *The Stanford Encyclopedia of Philosophy*, summer 2016 ed., edited by E. N. Zalta. http://plato.stanford.edu/archives/sum2016/entries/cognition-animal/.

Arvan, M. (2013). "A New Theory of Free Will." *Philosophical Forum* 44(1): 1–48.

Ayer, A. J. (1954). "Freedom and Necessity." In *Philosophical Essays*, 271–284. London: Macmillan.

Balaguer, M. (2009). *Free Will as an Open Scientific Problem*. Cambridge, MA: MIT Press.

Baumeister, R. F., E. J. Masicampo, and C. N. DeWall (2009). "Prosocial Benefits of Feeling Free: Disbelief in Free Will Increases Aggression and Reduces Helpfulness." *Personality and Social Psychology Bulletin* 35(2): 260–268.

Beckermann, A., H. Flohr, and J. Kim, eds. (1992). *Emergence or Reduction? Essays on the Prospects of Nonreductive Physicalism*. Berlin: De Gruyter.

Beebee, H., and A. Mele (2002). "Humean Compatibilism." *Mind* 111(442): 201–223.

Belnap, N. D. (2005). "Branching Histories Approach to Indeterminism and Free Will." In *Truth and Probability: Essays in Honour of Hugues Leblanc*, edited by B. Brown and F. Lepage, 197–211. London: College Publications.

Bennett, M., D. Dennett, P. Hacker, and J. Searle (2007). *Neuroscience and Philosophy: Brain, Mind, and Language*. New York: Columbia University Press.

Bernstein, S., and J. Wilson (2016). “Free Will and Mental Quausation.” *Journal of the American Philosophical Association* 2(2): 310–331.

Berofsky, B. (2011). “Compatibilism without Frankfurt: Dispositional Analyses of Free Will.” In *Oxford Handbook of Free Will,* 2nd ed., edited by R. Kane, 153–174. Oxford: Oxford University Press.

Berofsky, B. (2012). *Nature’s Challenge to Free Will.* Oxford: Oxford University Press.

Bianconi, E., A. Piovesan, F. Facchin, A. Beraudi, R. Casadei, F. Frabetti, L. Vitale, M. C. Pelleri, S. Tassani, F. Piva, S. Perez-Amodio, P. Strippoli, and S. Canaider (2013). “An Estimation of the Number of Cells in the Human Body.” *Annals of Human Biology* 40(6): 463–471.

Bishop, R. C. (2002). “Chaos, Indeterminism, and Free Will.” In *The Oxford Handbook of Free Will,* 1st ed., edited by R. Kane, 111–124. Oxford: Oxford University Press.

Block, N. (2003). “Do Causal Powers Drain Away?” *Philosophy and Phenomenological Research* 67(1): 133–150.

Bloom, P. (2006). “My Brain Made Me Do It.” *Journal of Cognition and Culture* 6(1–2): 209–214.

Boden, M. A. (2016). *AI: Its Nature and Future.* Oxford: Oxford University Press.

Bourget, D., and D. J. Chalmers (2014). “What Do Philosophers Believe?” *Philosophical Studies* 170(3): 465–500.

Brass, M., and P. Haggard (2007). “To Do or Not to Do: The Neural Signature of Self-Control.” *Journal of Neuroscience* 27(34): 9141–9145.

Bratman, M. E. (1987). *Intention, Plans, and Practical Reason.* Cambridge, MA: Harvard University Press.

Buckholtz, J. W., M. T. Treadway, R. L. Cowan, N. D. Woodward, R. Li, M. S. Ansari, R. M. Baldwin, A. N. Schwartzman, E. S. Shelby, C. E. Smith, R. M. Kessler, and D. H. Zald (2010). “Dopaminergic Network Differences in Human Impulsivity.” *Science* 329(5991): 532.

Butterfield, J. (2012). “Laws, Causation and Dynamics at Different Levels.” *Interface Focus* 2(1): 101–114.

Camerer, C. F., G. Loewenstein, and M. Rabin, eds. (2004). *Advances in Behavioral Economics.* Princeton, NJ: Princeton University Press.

Campbell, J. (2010). “Control Variables and Mental Causation.” *Proceedings of the Aristotelian Society* 110: 15–30.

Carroll, S. (2011). “Free Will Is as Real as Baseball.” *Discover,* 13 July. http://blogs.discovermagazine.com/cosmicvariance/2011/07/13/free-will-is-as-real-as-baseball.

Carroll, S. (2016). *The Big Picture: On the Origins of Life, Meaning, and the Universe Itself.* New York: Dutton.

Carter, I. (2012). "Positive and Negative Liberty." In *The Stanford Encyclopedia of Philosophy,* spring ed., edited by E. N. Zalta. http://plato.stanford.edu/archives/spr2012/entries/liberty-positive-negative/.

Cartwright, N. (1979). "Causal Laws and Effective Strategies." *Noûs* 13(4): 419–437.

Chakravartty, A. (2014). "Scientific Realism." In *The Stanford Encyclopedia of Philosophy,* spring ed., edited by E. N. Zalta. http://plato.stanford.edu/archives/spr2014/entries/scientific-realism/.

Chernyak, N., T. Kushnir, K. M. Sullivan, and Q. Wang (2013). "A Comparison of American and Nepalese Children's Concepts of Freedom of Choice and Social Constraint." *Cognitive Science* 37(7): 1343–1355.

Churchland, P. M. (1981). "Eliminative Materialism and the Propositional Attitudes." *Journal of Philosophy* 78(2): 67–90.

Churchland, P. S. (1986). *Neurophilosophy: Toward a Unified Science of the Mind-Brain.* Cambridge, MA: MIT Press.

Clarke, R. (2009). "Dispositions, Abilities to Act, and Free Will: The New Dispositionalism." *Mind* 118(470): 323–351.

Copp, D. (2006). "On the Agency of Certain Collective Entities: An Argument from 'Normative Autonomy.'" *Midwest Studies in Philosophy* 30(1): 194–221.

Cook, G. (2011). "Neuroscience Challenges Old Ideas about Free Will: Celebrated Neuroscientist Michael S. Gazzaniga Explains the New Science behind an Ancient Philosophical Question." *Scientific American,* 15 November. http://www.scientificamerican.com/article/free-will-and-the-brain-michael-gazzaniga-interview/.

Davidson, D. (1980). *Essays on Actions and Events.* Oxford: Clarendon.

Davidson, D. (2001). "Mental Events." In *Essays on Actions and Events,* 2nd ed., 207–227. Oxford: Clarendon.

Deery, O. (2015). "Why People Believe in Indeterminist Free Will." *Philosophical Studies* 172(8): 2033–2054.

Deery, O., and E. Nahmias (2017). "Defeating Manipulation Arguments: Interventionist Causation and Compatibilist Sourcehood." *Philosophical Studies* 174(5): 1255–1276.

Dennett, D. (1984). *Elbow Room: The Varieties of Free Will Worth Wanting.* Cambridge, MA: MIT Press.

Dennett, D. (1987). *The Intentional Stance.* Cambridge, MA: MIT Press.

Dennett, D. (2003). *Freedom Evolves.* London: Penguin.

Dennett, D. (2009). "Intentional Systems Theory." In *The Oxford Handbook of Philosophy of Mind,* edited by A. Beckermann, B. P. McLaughlin, and S. Walter, 339–350. Oxford: Oxford University Press.

Dennett, D., and J. Haugeland (1987). "Intentionality." In *The Oxford Companion to the Mind,* edited by R. L. Gregory, 383–386. Oxford: Oxford University Press.

DeWitt, S. J., S. Aslan, and F. M. Filbey (2014). "Adolescent Risk-Taking and Resting State Functional Connectivity." *Psychiatry Research: Neuroimaging* 222(3): 157–164.

Dietrich, F., and C. List (2016a). "Mentalism versus Behaviourism in Economics: A Philosophy-of-Science Perspective." *Economics and Philosophy* 32(2): 249–281.

Dietrich, F., and C. List (2016b). "Reason-Based Choice and Context-Dependence: An Explanatory Framework." *Economics and Philosophy* 32(2): 175–229.

Dretske, F. I. (1999). "Machines, Plants and Animals: The Origins of Agency." *Erkenntnis* 51(1): 523–535.

Dupré, J. (1993). *The Disorder of Things: Metaphysical Foundations of the Disunity of Science.* Cambridge, MA: Harvard University Press.

Ebstein, R. P., M. V. Monakhov, Y. Lu, Y. Jiang, P. S. Lai, and S. H. Chew (2015). "Association between the Dopamine D4 Receptor Gene Exon III Variable Number of Tandem Repeats and Political Attitudes in Female Han Chinese." *Proceedings of the Royal Society B: Biological Sciences* 282: 20151360.

Edwards, J. (1754). *A Careful and Strict Enquiry into the Modern Prevailing Notions of that Freedom of Will, Which is Supposed to be Essential to Moral Agency, Virtue and Vice, Reward and Punishment, Praise and Blame.* Boston: S. Kneeland.

Ellis, G. F. R., F. Noble, and T. O'Connor, eds. (2012). "Top-Down Causation." Themed issue of *Interface Focus* 2(1).

Elzein, N., and T. K. Pernu (2017). "Supervenient Freedom and the Free Will Deadlock." *Disputatio* 9(45): 219–243.

England, D. C. (n.d.). "If You Were Drunk at the Time, Can You Be Convicted?" Criminal Defense Lawyer, https://www.criminaldefenselawyer.com/resources/does-being-drunk-mean-you-cant-be-convicted-a-crime.h.

Erskine, T. (2001). "Assigning Responsibilities to Institutional Moral Agents: The Case of States and Quasi-States." *Ethics and International Affairs* 15(2): 67–85.

Fara, M. (2008). "Masked Abilities and Compatibilism." *Mind* 117(468): 843–865.

Field, H. (2003). "Causation in a Physical World." In *The Oxford Handbook of Metaphysics,* edited by M. J. Loux and D. W. Zimmerman. 435–460. Oxford: Oxford University Press.

Fine, A. (1984). "The Natural Ontological Attitude." In *Scientific Realism,* edited by J. Leplin, 83–107. Berkeley: University of California Press.

Fischer, J. M., R. Kane, D. Pereboom, and M. Vargas (2007). *Four Views on Free Will.* Oxford: Blackwell.

Floridi, L. (2008). "The Method of Levels of Abstraction." *Minds and Machines* 18(3): 303–329.

Floridi, L. (2011). *The Philosophy of Information.* Oxford: Oxford University Press.

Fodor, J. (1974). "Special Sciences (or: The Disunity of Science as a Working Hypothesis)." *Synthese* 28(2): 97–115.

Frankfurt, H. (1969). "Alternate Possibilities and Moral Responsibility." *Journal of Philosophy* 66: 829–839.

Freedom House (2015). "Freedom in the World 2015." https://freedomhouse.org/report/freedom-world/freedom-world-2015.

French, P. A. (1984). *Collective and Corporate Responsibility.* New York: Columbia University Press.

Frigg, R., and C. Hoefer (2015). "The Best Humean System for Statistical Mechanics." *Erkenntnis* 80(3): 551–574.

Garrett, D. (2009). "Hume." In *The Oxford Handbook of Causation,* edited by H. Beebee, C. Hitchcock, and P. Menzies, 53–71. Oxford: Oxford University Press.

Gazzaniga, M. (2011). *Who's in Charge: Free Will and the Science of the Brain.* New York: HarperCollins.

Gilboa, I. (2007). "Free Will: A Rational Illusion." Working paper, Tel Aviv University.

Glynn, L. (2010). "Deterministic Chance." *British Journal for the Philosophy of Science* 61(1): 51–80.

Green, T. A. (2014). *Freedom and Criminal Responsibility in American Legal Thought.* New York: Cambridge University Press.

Greene, J. D., R. B. Sommerville, L. E. Nystrom, J. M. Darley, and J. D. Cohen (2001). "An fMRI Investigation of Emotional Engagement in Moral Judgment." *Science* 293(5537): 2105–2108.

Habermas, J. (2007). "The Language Game of Responsible Agency and the Problem of Free Will: How Can Epistemic Dualism Be Reconciled with Ontological Monism?" *Philosophical Explorations* 10(1): 13–50.

Haggard, P., S. Clark, and J. Kalogeras (2002). "Voluntary Action and Conscious Awareness." *Nature Neuroscience* 5(4): 382–385.

Hall, N. (2004). "Two Concepts of Causation." In *Causation and Counterfactuals,* edited by J. Collins, N. Hall, and L. A. Paul, 225–276. Cambridge, MA: MIT Press.

Halpern, J. Y., and J. Pearl (2005a). "Causes and Explanations: A Structural-Model Approach. Part I: Causes." *British Journal for the Philosophy of Science* 56(4): 843–887.

Halpern, J. Y., and J. Pearl (2005b). "Causes and Explanations: A Structural-Model Approach. Part II: Explanations." *British Journal for the Philosophy of Science* 56(4): 889–911.

Harari, Y. N. (2016). "Yuval Noah Harari on Big Data, Google and the End of Free Will." *Financial Times,* 26 August 2016.

Harmon, K. (2010). "Dopamine Determines Impulsive Behavior." *Scientific American,* 29 July. http://www.scientificamerican.com/article/dopamine-impulsive-addiction/.

Harris, S. (2012a). *Free Will.* New York: Simon and Schuster.

Harris, S. (2012b). "Life without Free Will," September 9. Blog post. http://www.samharris.org/blog/item/life-without-free-will.

Haslanger, S. (2015). "What Is a (Social) Structural Explanation?" *Philosophical Studies* 173(1): 113–130.

Hatemi, P. K., N. A. Gillespie, L. J. Eaves, B. S. Maher, B. T. Webb, A. C. Heath, S. E. Medland, D. C. Smyth, H. N. Beeby, S. D. Gordon, G. W. Montgomery, G. Zhu, E. M. Byrne, and N. G. Martin (2011). "A Genome-Wide Analysis of Liberal and Conservative Political Attitudes." *Journal of Politics* 73(1): 271–285.

Haynes, J.-D., K. Sakai, G. Rees, S. Gilbert, C. Frith, and R. E. Passingham (2007). "Reading Hidden Intentions in the Human Brain." *Current Biology* 17: 323–328.

Heider, F., and M. Simmel (1944). "An Experimental Study of Apparent Behavior." *American Journal of Psychology* 157: 243–259.

Himmelreich, J. (2015). "Agency as Difference-Making: Causal Foundations of Moral Responsibility." PhD diss., London School of Economics and Political Science.

Hodgson, D. (2002). "Quantum Physics, Consciousness, and Free Will." In *The Oxford Handbook of Free Will,* 1st ed., edited by R. Kane, 85–110. Oxford: Oxford University Press.

Hoefer, C. (2002). "Freedom from the Inside Out." *Royal Institute of Philosophy Supplements* 50: 201–222.

Honderich, T. (2002). "Determinism as True, Both Compatibilism and Incompatibilism as False, and the Real Problem." In *The Oxford Handbook of Free Will,* 1st ed., edited by R. Kane, 461–476. Oxford: Oxford University Press.

Horgan, T. (1993). "From Supervenience to Superdupervenience: Meeting the Demands of a Material World." *Mind* 102(408): 555–586.

Hume, D. ([1739–40] 1888). *A Treatise of Human Nature.* Edited by L. A. Selby-Bigge. Oxford: Clarendon Press.

Hurley, S. (1999). "Responsibility, Reason, and Irrelevant Alternatives." *Philosophy and Public Affairs* 28(3): 205–241.

Ismael, J. T. (2009). "Probability in Deterministic Physics." *Journal of Philosophy* 106(2): 89–108.

Ismael, J. T. (2013). "Causation, Free Will, and Naturalism." In *Scientific Metaphysics,* edited by D. Ross, J. Ladyman, and H. Kincaid, 208–235. Oxford: Oxford University Press.

Ismael, J. T. (2016). *How Physics Makes Us Free.* New York: Oxford University Press.

Jackson, F., and P. Pettit (1990). "Program Explanation: A General Perspective." *Analysis* 50(2): 107–117.

Jardim-Messeder, D., K. Lambert, S. Noctor, F. M. Pestana, M. E. de Castro Leal, M. F. Bertelsen, A. N. Alagaili, O. B. Mohammad, P. R. Manger, and S. Herculano-Houzel (2017). "Dogs Have the Most Neurons, Though Not the Largest Brain: Trade-Off between Body Mass and Number of Neurons in the Cerebral Cortex of Large Carnivoran Species." *Frontiers in Neuroanatomy* 11: 118.

Kahneman, D. (2011). *Thinking, Fast and Slow.* New York: Farrar, Straus and Giroux.

Kane, R. (1998). *The Significance of Free Will.* Oxford: Oxford University Press.

Kane, R. (1999). "Responsibility, Luck, and Chance: Reflections on Free Will and Indeterminism." *Journal of Philosophy* 96(5): 217–240.

Kane, R., ed. (2002). *The Oxford Handbook of Free Will.* Oxford: Oxford University Press.

Kawohl, W., and E. Habermeyer (2007). "Free Will: Reconciling German Civil Law with Libet's Neurophysiological Studies on the Readiness Potential." *Behavioral Sciences and the Law* 25(2): 309–320.

Kenny, A. (1978). *Freewill and Responsibility.* London: Routledge.

Kim, J. (1984). "Concepts of Supervenience." *Philosophy and Phenomenological Research* 45(2): 153–176.

Kim, J. (1987). "'Strong' and 'Global' Supervenience Revisited." *Philosophy and Phenomenological Research* 48(2): 315–326.

Kim, J. (1993). "The Nonreductivist's Troubles with Mental Causation." In *Supervenience and Mind: Selected Philosophical Essays,* 336–357. Cambridge: Cambridge University Press.

Kim, J. (1998). *Mind in a Physical World: An Essay on the Mind-Body Problem and Mental Causation.* Cambridge, MA: MIT Press.

Kim, J. (2002). "The Layered Model: Metaphysical Considerations." *Philosophical Explorations* 5(1): 2–20.

Kim J. (2005). *Physicalism, or Something Near Enough.* Princeton, NJ: Princeton University Press.

Knobe, J. (2003). "Intentional Action and Side Effects in Ordinary Language." *Analysis* 63(279): 190–194.

Kramer, M. (2003). *The Quality of Freedom.* Oxford: Oxford University Press.

Kratzer, A. (1977). "What 'Must' and 'Can' Must and Can Mean." *Linguistics and Philosophy* 1: 335–355.

Kushnir, T., A. Gopnik, N. Chernyak, E. Seiver, and H. M. Wellman (2015). "Developing Intuitions about Free Will between Ages Four and Six." *Cognition* 138: 79–101.

Levin, J. (2016). "Review of Polger and Shapiro, *The Multiple Realization Book.*" *Notre Dame Philosophical Reviews,* 12 December. https://ndpr.nd.edu/news/the-multiple-realization-book/.

Lewis, D. (1973). "Causation." *Journal of Philosophy* 70(17): 556–567.

Lewis, D. (1986). *The Plurality of Worlds.* Oxford: Blackwell.

Libet, B., C. A. Gleason, E. W. Wright, and D. K. Pearl (1983). "Time of Conscious Intention to Act in Relation to Onset of Cerebral Activity (Readiness-Potential): The Unconscious Initiation of a Freely Voluntary Act." *Brain* 106: 623–642.

List, C. (2012). "Free Will in a Deterministic World?" Forum for European Philosophy, London School of Economics and Political Science, 4 December. Podcast. http://www.lse.ac.uk/lse-player?id=1678.

List, C. (2014). "Free Will, Determinism, and the Possibility of Doing Otherwise." *Noûs* 48(1): 156–178.

List, C. (2015). "What's Wrong with the Consequence Argument: In Defence of Compatibilist Libertarianism." Working paper, London School of Economics and Political Science.

List, C. (2018). "Levels: Descriptive, Explanatory, and Ontological." *Noûs,* forthcoming.

List, C., and P. Menzies (2009). "Non-reductive Physicalism and the Limits of the Exclusion Principle." *Journal of Philosophy* 106(9): 475–502.

List, C., and P. Menzies (2017). "My Brain Made Me Do It: The Exclusion Argument against Free Will, and What's Wrong with It." In *Making a Difference: Essays on the Philosophy of Causation,* edited by H. Beebee, C. Hitchcock, and H. Price, 269–285. Oxford: Oxford University Press.

List, C., and P. Pettit (2011). *Group Agency: The Possibility, Design, and Status of Corporate Agents.* Oxford: Oxford University Press.

List, C., and M. Pivato (2015). "Emergent Chance." *Philosophical Review* 124(1): 119–152.

List, C., and M. Pivato (2016). "Dynamic and Stochastic Systems as a Framework for Metaphysics and the Philosophy of Science." Manuscript, 23 April. http://philsci-archive.pitt.edu/12371/.

List, C., and W. Rabinowicz (2014). "Two Intuitions about Free Will: Alternative Possibilities and Intentional Endorsement." *Philosophical Perspectives* 28: 155–172.

List, C., and K. Spiekermann (2013). "Methodological Individualism and Holism in Political Science: A Reconciliation." *American Political Science Review* 107(4): 629–643.

List, C., and L. Valentini (2016). "Freedom as Independence." *Ethics* 126(4): 1043–1074.

Loewer, B. (2001). "Determinism and Chance." *Studies in History and Philosophy of Modern Physics* 32B(4): 609–620.

Mackintosh, N. (2011). "My Brain Made Me Do It." *New Scientist* 212(2843): 26–27.

Maier, J. (2015). "The Agentive Modalities." *Philosophy and Phenomenological Research* 90(1): 113–134.

Manafu, A. (2015). "A Novel Approach to Emergence in Chemistry." In *Philosophy of Chemistry: Growth of a New Discipline,* edited by E. Scerri and L. McIntyre, 39–55. Heidelberg: Springer.

McLaughlin, B., and K. Bennett (2014). "Supervenience." In *The Stanford Encyclopedia of Philosophy,* spring ed., edited by E. N. Zalta. http://plato.stanford.edu/archives/spr2014/entries/supervenience/.

Melden, A. I. (1961). *Free Action.* London: Routledge.

Mele, A. R. (2014). *Free: Why Science Hasn't Disproved Free Will.* Oxford: Oxford University Press.

Mele, A. R. (2017). *Aspects of Agency: Decisions, Abilities, Explanations, and Free Will.* Oxford: Oxford University Press.

Menzies, P. (2014). "Counterfactual Theories of Causation." In *The Stanford Encyclopedia of Philosophy,* spring ed., edited by E. N. Zalta. https://plato.stanford.edu/archives/spr2014/entries/causation-counterfactual/.

Menzies, P., and C. List (2010). "The Causal Autonomy of the Special Sciences." In *Emergence in Mind,* edited by C. Mcdonald and G. Mcdonald, 108–128. Oxford: Oxford University Press.

Menzies, P., and H. Price (1993). "Causation as a Secondary Quality." *British Journal for the Philosophy of Science* 44(2): 187–203.

Moore, G. E. (1912). *Ethics.* Oxford: Oxford University Press.

Müller, T., and T. Placek (2018). "Defining Determinism." *British Journal for the Philosophy of Science* 69(1): 215–252.

Musser, G. (2015). "Is the Cosmos Random?" *Scientific American,* September, 88–93.

Nahmias, E. (2010). "Scientific Challenges to Free Will." In *A Companion to the Philosophy of Action,* edited by T. O'Connor and C. Sandis, 345–356. Oxford: Wiley-Blackwell.

Nahmias, E. (2014). "Is Free Will an Illusion? Confronting Challenges from the Modern Mind Sciences." In *Moral Psychology.* Vol. 4, *Freedom and Responsibility,* edited by W. Sinnott-Armstrong, 1–57. Cambridge, MA: MIT Press.

Nahmias, E. (2015). "Why We Have Free Will." *Scientific American,* January, 77–79.

Nahmias, E., J. Shepard, and S. Reuter (2014). "It's OK if 'My Brain Made Me Do It': People's Intuitions about Free Will and Neuroscientific Prediction." *Cognition* 133(2): 502–516.

Nichols, S. (2006). "Folk Intuitions on Free Will." *Journal of Cognition and Culture* 6(1–2): 57–86.

Norton, J. (2003). "Causation as Folk Science." *Philosophers' Imprint* 3(4): 1–22.

O'Connor, T., and C. Franklin (2018). "Free Will." In *The Stanford Encyclopedia of Philosophy,* fall ed., edited by E. N. Zalta. https://plato.stanford.edu/archives/fall2018/entries/freewill/.

O'Leary-Hawthorne, J., and P. Pettit (1996). "Strategies for Free Will Compatibilists." *Analysis* 56(4): 191–201.

Oppenheim, P., and H. Putnam (1958). "Unity of Science as a Working Hypothesis." *Minnesota Studies in the Philosophy of Science* 2: 3–36.

Owens, D. (1989). "Levels of Explanation." *Mind* 98(389): 59–79.

Pearl, J. (2000). *Causality: Models, Reasoning, and Inference.* New York: Cambridge University Press.

Penrose, R. (1994). *Shadows of the Mind: A Search for the Missing Science of Consciousness.* Oxford: Oxford University Press.

Pereboom, D. (2001). *Living without Free Will.* Cambridge: Cambridge University Press.

Pettit, P. (1997). *Republicanism: A Theory of Freedom and Government.* Oxford: Oxford University Press.

Pettit, P. (2002). *A Theory of Freedom: From the Psychology to the Politics of Agency.* New York: Oxford University Press.

Pettit, P. (2007). "Responsibility Incorporated." *Ethics* 117(2): 171–201.

Pettit, P., and M. Smith (1996). "Freedom in Belief and Desire." *Journal of Philosophy* 93(9): 429–449.

Polger, T. W., and L. A. Shapiro (2016). *The Multiple Realization Book.* Oxford: Oxford University Press.

Putnam, H. (1975a). *Mathematics, Matter and Method.* Cambridge: Cambridge University Press.

Putnam, H. (1975b). "Philosophy and Our Mental Life." In *Philosophical Papers.* Vol. 2, *Mind, Language and Reality,* 291–303. Cambridge: Cambridge University Press.

Pyysiäinen, I. (2009). *Supernatural Agents: Why We Believe in Souls, Gods, and Buddhas.* Oxford: Oxford University Press.

Quian Quiroga, R., I. Fried, and C. Koch (2013). "Brain Cells for Grandmother." *Scientific American,* February 2013, 31–35.

Quine, W. V. (1977). *Ontological Relativity and Other Essays.* New York: Columbia University Press.

Raatikainen, P. (2010). "Causation, Exclusion, and the Special Sciences." *Erkenntnis* 73(3): 349–363.

Ramsey, W. (2013). "Eliminative Materialism." In *The Stanford Encyclopedia of Philosophy,* summer ed., edited by E. N. Zalta. http://plato.stanford.edu/archives/sum2013/entries/materialism-eliminative/.

Rickless, S. C. (2000). "Locke on the Freedom to Will." *Locke Newsletter* 31: 43–67.

Roskies, A. L. (2006). "Neuroscientific Challenges to Free Will and Responsibility." *Trends in Cognitive Science* 10(9): 419–423.

Roskies, A. L. (2010). "Why Libet's Studies Don't Pose a Threat to Free Will." In *Conscious Will and Responsibility: A Tribute to Benjamin Libet,* edited by W. Sinnott-Armstrong and L. Nadel, 11–22. Oxford: Oxford University Press.

Roskies, A. L. (2012). "Don't Panic: Self-Authorship without Obscure Metaphysics." *Philosophical Perspectives* 26: 323–342.

Rovelli, C. (2013). "Free Will, Determinism, Quantum Theory and Statistical Fluctuations: A Physicist's Take." *Edge,* 8 August. https://www.edge.org/conversation/free-will-determinism-quantum-theory-and-statistical-fluctuations-a-physicists-take.

Russell, B. (1913). "On the Notion of Cause." *Proceedings of the Aristotelian Society* 13(1): 1–26.

Russell, S. J., and P. Norvig (2009). *Artificial Intelligence: A Modern Approach.* 3rd ed. Upper Saddle River, NJ: Prentice Hall.

Rutledge, R. B., N. Skandali, P. Dayan, and R. J. Dolan (2015). "Dopaminergic Modulation of Decision Making and Subjective Well-Being." *Journal of Neuroscience* 35(27): 9811–9822.

Sarkissian, H., A. Chatterjee, F. de Brigard, J. Knobe, S. Nichols, and S. Sirker (2010). "Is Belief in Free Will a Cultural Universal?" *Mind and Language* 25(3): 346–358.

Sartre, J.-P. (1992). *Being and Nothingness.* Translated by H. E. Barnes. New York: Washington Square Press.

Schaffer, J. (2003). "Is There a Fundamental Level?" *Noûs* 37(3): 498–517.

Schaffer, J. (2009). "On What Grounds What." In *Metametaphysics: New Essays on the Foundations of Ontology,* edited by D. Chalmers, D. Manley, and R. Wasserman, 347–383. Oxford: Oxford University Press.

Searle, J. R. (1979). "What Is an Intentional State?" *Mind* 88(349): 74–92.

Searle, J. R. (1983). *Intentionality: An Essay in the Philosophy of Mind.* Cambridge: Cambridge University Press.

Seebass, G. (1993a). "Freiheit und Determinismus (Teil 1)." *Zeitschrift für philosophische Forschung* 47: 1–22.

Seebass, G. (1993b). "Freiheit und Determinismus (Teil 2)." *Zeitschrift für philosophische Forschung* 47: 223–245.

Settle, J. E., C. T. Dawes, N A. Christakis, and J. H. Fowler (2010). "Friendships Moderate an Association between a Dopamine Gene Variant and Political Ideology." *Journal of Politics* 72(4): 1189–1198.

Sewell, W. H. (1992). "A Theory of Structure: Duality, Agency, and Transformation." *American Journal of Sociology* 98(1): 1–29.

Shariff, A. F., J. D. Greene, J. C. Karremans, J. B. Luguri, C. J. Clark, J. W. Schooler, R. F. Baumeister, and K. D. Vohs (2014). "Free Will and Punishment: A Mechanistic View of Human Nature Reduces Retribution." *Psychological Science* 25(8): 1563–1570.

Shermer, M. (2009). "Why People Believe Invisible Agents Control the World." *Scientific American,* 1 June. http://www.scientificamerican.com/article/skeptic-agenticity/.

Shermer, M. (2012). "How Free Will Collides with Unconscious Impulses." *Scientific American,* 1 August. https://www.scientificamerican.com/article/how-free-will-collides-with-unconscious-impulses/.

Siderits, M. (2008). "Paleo-Compatibilism and Buddhist Reductionism." *Sophia* 47(1): 29–42.

Siewert, C. (2017). "Consciousness and Intentionality." In *The Stanford Encyclopedia of Philosophy,* spring ed., edited by E. N. Zalta. https://plato.stanford.edu/archives/spr2017/entries/consciousness-intentionality/.

Singer, W. (2007). "Binding by Synchrony." *Scholarpedia* 2(12): 1657. https://doi.org/10.4249/scholarpedia.1657.

Smilansky, S. (2000). *Free Will and Illusion.* Oxford: Oxford University Press.

Smith, L. (2007). *Chaos: A Very Short Introduction.* Oxford: Oxford University Press.

Smith, M. (2003). "Rational Capacities, or: How to Distinguish Recklessness, Weakness, and Compulsion." In *Weakness of Will and Practical Irrationality,* edited by S. Stroud and C. Tappolet, 17–38. Oxford: Oxford University Press.

Snidal, D. (1985). "The Game Theory of International Politics." *World Politics* 38(1): 25–57.

Sober, E. (2010). "Evolutionary Theory and the Reality of Macro-Probabilities." In *The Place of Probability in Science,* edited by E. Eells and J. H. Fetzer, 133–161. Heidelberg: Springer.

Sternberg, E. J. (2010). *My Brain Made Me Do It: The Rise of Neuroscience and the Threat to Moral Responsibility.* Amherst, New York: Prometheus.

Steward, H. (2009). "Animal Agency." *Inquiry* 52(3): 217–231.

Steward, H. (2012). *A Metaphysics for Freedom.* Oxford: Oxford University Press.

Steward, H. (2015). "Do Animals Have Free Will?" *Philosophers' Magazine* 68: 43–48.

Strawson, G. (1994). "The Impossibility of Moral Responsibility." *Philosophical Studies* 75(1–2): 5–24.

Strevens, M. (2011). "Probability out of Determinism." In *Probabilities in Physics,* edited by C. Beisbart and S. Hartmann, 339–364. Oxford: Oxford University Press.

Szalavitz, M. (2012). "My Brain Made Me Do It: Psychopaths and Free Will." *Time,* 17 August. http://healthland.time.com/2012/08/17/my-brain-made-me-do-it-psychopaths-and-free-will/.

Tadros, V. (2016). *Wrongs and Crimes.* Oxford: Oxford University Press.

Taylor, C., and D. Dennett (2002). "Who's Afraid of Determinism? Rethinking Causes and Possibilities." In *The Oxford Handbook of Free Will,* 1st ed., edited by R. Kane, 257–277. Oxford: Oxford University Press.

Tollefsen, D. P. (2015). *Groups as Agents.* Cambridge: Polity.

Tononi, G., and C. Koch (2015). "Consciousness: Here, There and Everywhere?" *Philosophical Transactions of the Royal Society B: Biological Sciences* 370: 20140167.

Treisman, A. (1996). "The Binding Problem." *Current Opinion in Neurobiology* 6(2): 171–178.

Uithol, S., D. C. Burnston, and P. Haselager (2014). "Why We May Not Find Intentions in the Brain." *Neuropsychologia* 56: 129–139.

Van Fraassen, B. C. (1980). *The Scientific Image.* Oxford: Oxford University Press.

Van Gelder, T. (1995). "What Might Cognition Be, if Not Computation?" *Journal of Philosophy* 92(7): 345–381.

Van Inwagen, P. (1975). "The Incompatibility of Free Will and Determinism." *Philosophical Studies* 27(3): 185–199.

Van Inwagen, P. (1983). *An Essay on Free Will.* Oxford: Oxford University Press.

Vihvelin, K. (2000). "Libertarian Compatibilism." *Philosophical Perspectives* 14: 139–166.

Vihvelin, K. (2004). "Free Will Demystified: A Dispositional Account." *Philosophical Topics* 32(1–2): 427–450.

Vihvelin, K. (2013). *Causes, Laws, and Free Will: Why Determinism Doesn't Matter.* Oxford: Oxford University Press.

Vohs, K. D., and J. W. Schooler (2008). "The Value of Believing in Free Will: Encouraging a Belief in Determinism Increases Cheating." *Psychological Science* 19(1): 49–54.

Von Plato, J. (1982). "Probability and Determinism." *Philosophy of Science* 49(1): 51–66.

Walter, H. (2001). *Neurophilosophy of Free Will: From Libertarian Illusions to a Concept of Natural Autonomy.* Cambridge, MA: MIT Press.

Waltz, K. (1979). *Theory of International Politics.* Long Grove, IL: Waveland.

Wegner, D. M. (2002). *The Illusion of Conscious Will.* Cambridge, MA: MIT Press.

Werndl, C. (2009). "Are Deterministic Descriptions and Indeterministic Descriptions Observationally Equivalent?" *Studies in History and Philosophy of Science Part B* 40(3): 232–242.

Whittle, A. (2010). "Dispositional Abilities." *Philosophers' Imprint* 10(12): 1–23.

Winfield, A. F. T. (2012). *Robotics: A Very Short Introduction.* Oxford: Oxford University Press.

Wood, A. W. (1999). *Kant's Ethical Thought.* Cambridge: Cambridge University Press.

Woodward, J. (2003). *Making Things Happen: A Theory of Causal Explanation.* New York: Oxford University Press.

Woodward, J. (2008). "Mental Causation and Neural Mechanisms." In *Being Reduced: New Essays on Reduction, Explanation, and Causation,* edited by J. Hohwy and J. Kallestrup, 218–262. Oxford: Oxford University Press.

Woodward, J. (2009). "Agency and Interventionist Theories." In *The Oxford Handbook of Causation,* edited by H. Beebee, C. Hitchcock, and P. Menzies, 234–262. Oxford: Oxford University Press.

Yalowitz, S. (2014). "Anomalous Monism." In *The Stanford Encyclopedia of Philosophy,* winter ed., edited by E. N. Zalta. https://plato.stanford.edu/archives/win2014/entries/anomalous-monism/.

Yoshimi, J. (2012). "Supervenience, Dynamical Systems Theory, and Non-reductive Physicalism." *British Journal for the Philosophy of Science* 63(2): 373–398.

Zuckerman, M., and M. D. Kuhlman (2000). "Personality and Risk-Taking: Common Biosocial Factors." *Journal of Personality* 68(6): 999–1029.

Acknowledgments

I would like to express my gratitude to the large number of people who have helped me with the work presented in this book, through their advice and feedback, critical comments and questions, and broader conversations and encouragement. Although it is impossible to name all of them here, I would like to mention some in particular.

First of all, I am greatly indebted to my collaborators on projects related to the topic of this book: the late Peter Menzies, with whom I have worked on mental causation and its relevance to free will and who has been a key source of inspiration; Wlodek Rabinowicz, with whom I have worked on the relationship between free will and intentional endorsement and who has been a longstanding philosophical interlocutor; Philip Pettit, with whom I have worked on intentional agency in individuals and groups and who has been my closest academic mentor; Franz Dietrich, with whom I have worked on rationality and reasons for choice and discussed just about every philosophical topic under the sun; and Marcus Pivato, with whom I have worked on determinism, indeterminism, and emergent phenomena and who has been an unfailing source of insight. All of them have influenced my work in many more ways than I can properly acknowledge here, and I am most grateful for their friendship and wisdom.

Secondly, several people have read drafts of this book in its entirety, or significant portions of it, and I have greatly benefitted from their generous comments and feedback. I would like to mention David Axelsen, Jonathan Birch, Henrik Dahlquist, Roman Frigg, Mathias Koenig-Archibugi, Klaus Jürgen

List, Anna Mahtani, Mike Otsuka, Laura Valentini, and Peter Wilson, as well as the participants in a series of seminars on free will that I gave at the London School of Economics and Political Science in the autumn of 2017, including Will Bosworth, Federico Brandmayr, Chloé de Canson, Margherita Harris, Todd Karhu, Sophie Kikkert, David Kinney, Ko-Hung Kuan, Bertram O'Brien, Joe Roussos, Aron Vallinder, and Cecily Whiteley. I have also benefitted from the helpful comments that I have received from my editor, Ian Malcolm, and the reviewers for Harvard University Press, including Alfred Mele. In the early stages of this book project, I received valuable research assistance from Bharath Palle and Colin Reeves. I would also like to thank Robert Kane, whose advice and encouragement have made a big difference to my free-will project as a whole.

In addition to the people mentioned so far, several others have kindly given me comments and feedback on some aspects of my work on free will, sometimes in writing, sometimes in conversations, and sometimes at seminars or conferences. They include Matthew Adler, Arif Ahmed, Miri Albahari, Jason Alexander, Maria Alvarez, Alexandru Baltag, Tim Bayne, Helen Beebee, Claus Beisbart, Bernard Berofsky, Luc Bovens, Richard Bradley, Liam Kofi Bright, Jan Broersen, John Broome, Eva Buddeberg, Susanne Burri, Jeremy Butterfield, Rosa Cao, Sean Carroll, David Chalmers, Chlump Chatkupt, Larissa Conradt, John Danaher, Daniel Dennett, Cian Dorr, Hein Duijf, Benjamin Eva, Rainer Forst, Roman Frigg, Alexander Gebharter, Itzhak Gilboa, Natalie Gold, Francesco Guala, Alan Hájek, Daniel Harbour, Pim Haselager, John Hawthorne, Frank Hofmann, Richard Holton, Meir Hemmo, Johannes Himmelreich, George Huxford, John Hyman, Elizabeth Kamali, Christian Loew, John Maier, Joseph Mazor, Ariel Mendez, Silvia Milano, Thomas Müller, Kristina Musholt, George Musser, Eddy Nahmias, Daniel Nolan, Deren Olgun, Dan Osherson, Chris Ovenden, Ian Phillips, Hanna Pickard, Huw Price, Henry Richardson, Bryan Roberts, Asbjørn Aagaard Schmidt, Marek Sergot, Orly Shenker, Walter Sinnott-Armstrong, Sonja Smets, Kai Spiekermann, Reuben Stern, Daniel Stoljar, David Storrs-Fox, Victor Tadros, Giulio Tononi, Adrian Vermeule, Kadri Vihvelin, Kate Vredenburgh, Tobias Hansson Wahlberg, Ralph Weir, Ann Whittle, and John Wright.

I have given many talks and lectures on the themes discussed in this book, and although I cannot enumerate them all, I would like to record my thanks to the audiences who have engaged with my work. These include audiences at

Harvard Law School, King's College London, the London School of Economics and Political Science (LSE), Lund University, Stanford University, the Swedish Collegium for Advanced Study, the University of Bern, the University of Frankfurt, the University of Glasgow, the University of Luxembourg, the University of Manchester, the University of Oxford, the University of Paris (Sorbonne), the University of Salzburg, the University of Warwick, and Utrecht University.

I would further like to thank several institutions that have generously supported my work on this project. I am very grateful to the Leverhulme Trust for supporting me through a Leverhulme Major Research Fellowship on the project titled "Reasons, Decisions, and Intentional Agency." I thank the Department of Philosophy at the LSE and the LSE's Centre for Philosophy of Natural and Social Science for providing a wonderful workplace. I am particularly grateful to my friends and colleagues in the Choice Group and to my students at undergraduate and graduate levels from all over the world in both the philosophy and the political theory programmes. I also gratefully acknowledge support from the LSE's Department of Government, including financial support for a workshop on free will which I co-organized with my friend and colleague Kai Spiekermann in the summer of 2013.

I have had the privilege of spending two semesters in Sweden, at the Swedish Collegium for Advanced Study and the University of Uppsala (in the autumns of 2011 and 2014), a northern summer (or southern winter) at the Australian National University (in 2014), a semester at Harvard Law School (in the autumn of 2015), and part of a semester at the Forschungskolleg Humanwissenschaften of the Goethe-Universität, Frankfurt (in early 2018). I hugely benefitted from the exchange of ideas with colleagues and students at all of these places. Much of my time at Harvard and Frankfurt, in particular, was devoted to work on the present manuscript, and my visits there made a big difference to this book. I fondly remember my long work sessions on the Wingertsberg in Bad Homburg as I was completing the final chapters.

The present book builds on some of my earlier work on the topic of free will. In particular, Chapter 4 and the book's conclusion build on and expand ideas first published in "Free Will, Determinism, and the Possibility of Doing Otherwise," *Noûs* 48(1): 156–178, 2014 (doi: 10.1111/nous.12019). The book is also inspired by my joint work with Peter Menzies and Wlodek Rabinowicz, as cited in relevant places, and I would like to reiterate my intellectual gratitude to both of them.

My biggest thank you goes to my wife, Laura Valentini. She has not only been my greatest support, personal as well as intellectual, but she has also discussed the free-will problem with me countless times and given me the most astute advice. I am incredibly lucky to share my life with her. A huge thank you also to my parents, Mutti und Vati, on whom I have always been able to count and who instilled in me the intellectual curiosity that led me to philosophy. A big thank you, finally, to my parents-in-law, Mamma e Papà, who have been another great source of support (as well as of lovely Italian food during long visits to Italy!). I dedicate this book to Laura, Mutti, Vati, Mamma, and Papà.

Index

"Aboutness," 67–68, 121. *See also* Intentionality
Agency, intentional. *See* Intentional agency
Agency incompatibilism, 179n23
Agential determinism: definition of, 87; incompatibility with free will, 152; as independent of physical determinism, 89–97; inevitability and, 177–178n13; option sets and, 100; premises of, 86–88
Agential indeterminism: compatibility with physical determinism, 81, 89–96, 152; definition of, 89; human agency and, 97–103, 178–179n18; intentional endorsement and, 110–111, 180–181n34; randomness versus, 43, 107–111. *See also* Alternative possibilities
Agenticity, 170n3
Agentive modalities, 179n20
Alternative possibilities: case for realism about, 97–103; compatibility with ordinary language, 103–107; definition of, 23–24; higher-level nature of, 4–5, 150; imagined possibilities, 102–103; interpretations of, 81–86, 177n8; moral responsibility and, 23–24, 168–169n29; as prerequisite to free will, 16, 26, 79–81; randomness, 107–111. *See also* Agential indeterminism; Physical determinism
Alvarez, Maria, 168–169n29, 169n32
Amygdala, risk-taking behaviour and, 36
Andrews, Kristin, 174n13
Animals: free will in, 153–154, 185n8; intentional agency of, 58–62, 173–174n14, 173nn11–13
"Anomalism of the Mental" (Davidson), 178–179n18
Artificial intelligence, 156–157
Association, laws of, 130
Atoms, structure of, 125
Authorship of actions, 27, 108, 168–169n29
Automatism defense, 18

Balaguer, Mark, 12, 163–164n18
Beebee, Helen, 12
Behaviourism, 75
Belief-and-desire states: as attitudes towards meaningful contents, 53–54,

Belief-and-desire states (*continued*) 67; belief-desire-intention agent, 172n1; instrumental rationality, 51–52; intentional stance, 55–56, 154–155; neural encoding of, 36, 71; objects of, 172n4; obstacles to reduction, 67–68, 175n21
Bell's theorem, 42
Berofsky, Bernard, 12
Big Picture, The (Carroll), 164–165n20
Block, Ned, 124–125
Borges, Jorge Luis, 79
Bourget, David, 153
Brass, Marcel, 145
Bratman, Michael, 172n1
Burnston, Daniel, 170–171n8
Butterfield, Jeremy, 97

Campbell, John, 131, 138
"Can," semantics of, 103–107
"Can do otherwise." *See* Alternative possibilities
Capacity for free will, 27–29
Carroll, Sean, 164–165n20
Cartwright, Nancy, 129–130
Causal closure principle, 45, 119, 122–123, 139, 182n7
Causal control: causal exclusion argument, 44–45, 118–123, 182n7, 182n8, 184n33; causal source, 113–114; cause and effect, 114–118; definition of, 24–26; difference making, causation as, 131–140; folk science, causation as, 183n22; higher-level nature of, 4–5, 124–132; intentionality and, 25–26, 52–54, 169n33; laws of association and, 130; Libet experiments, 47, 121–122, 141–147; mental causation, 114, 118–123, 136–137, 150; as prerequisite to free will, 16, 113–114, 169n32; realism, case for, 132–140. *See also* Epiphenomenalism
Causal exclusion argument, 44–45; causal closure principle, 45, 119, 122–123, 139, 182n7; causal exclusion principle, 45, 119–120, 122–123, 139, 184n33; generalization of, 122–123; genuine causal overdetermination, 119, 182n8, 182n11; nonidentity principle, 120–121
"Causal laws," 127–128
"Causal oomph," 184n33
Causal reasoning, 114–118
Causal source, 113–114
Cause and effect, 114–118
CERN (Conseil Européen pour la Recherche Nucléaire), 102
Challenges to free will. *See* Determinism, physical; Epiphenomenalism; Radical materialism
Chalmers, David, 153
Chernyak, Nadia, 17
Churchland, Patricia, 32
Churchland, Paul, 32, 35–36
Classical mechanics, 55, 87, 126–127
"Clockwork universe" concept, 40
Collectives, agency in, 154–156
Compatibilism: Humean, 12, 115–117, 172n1, 185n3; libertarian, 12, 164n19; paleo-compatibilism, 163n17; premises of, 151–153. *See also* Compatibilist libertarianism
Compatibilist libertarianism, 162n10; definition of, 9–10; empirical premises, 10–11, 151–152; precursors to, 12; third-person perspective, 11; thought experiment, 157–158. *See also* Alternative

possibilities; Causal control; Intentional agency
Conditional interpretation of alternative possibilities, 81–86, 177n8
Conscious intention, neuroscientific studies of, 47–48, 64, 141–147
Consent, free will and, 18
Consequence Argument, 163n12
Constraints on action, 16–17, 20–21, 104, 127, 167–168n25
Constructive empiricism, 176n33
Consumer theory, intentional agency in, 32
Control, causal. *See* Causal control
Control variables, 138, 146
Copenhagen interpretation of quantum mechanics, 41, 97, 171n22, 176n1
Correlation, 129–130
"Could have done otherwise." *See* Alternative possibilities
Coyne, Jerry, 1
Criminal law, presupposition of free will in, 17–20, 23, 165n8, 166n10, 167n22

Darwin, Charles, 158
Davidson, Donald, 25, 178–179n18
Decision making: decision theory, 61–62; presupposition of free will in, 17–20, 165n8, 166n10, 167n22
Degree of free will, 29–30
Deliberation, presupposition of free will in, 17–20, 103, 165n8, 166n10, 167n22
Dennett, Daniel, 12, 24, 55–57, 154–155, 177
Design stance, 56
Desire. *See* Belief-and-desire states
Determinism, agential. *See* Agential determinism
Determinism, physical: agency incompatibilism, 179n23; compatibility with agential indeterminism, 89–97, 152; definition of, 2–3, 87; in Enlightenment worldview, 2, 38–40, 87; hidden-variables interpretations, 42; inevitability and, 177–178n13; Laplace's demon, 40; linking thesis, 88–89; physics, laws of, 39–41; premises of, 2–4, 26, 38–39, 79, 86–87, 149; probability and, 178n17; randomness, 107–111; randomness and, 43; thought experiment, 157–158. *See also* Alternative possibilities; Lower-level phenomenon
Deterministic histories: physical (lower-level), 93–95; psychological (higher-level), 94–97
Dietrich, Franz, 172n2
Difference making, causation as, 131–140
Dispositional interpretation of alternative possibilities, 81–86, 177n8
"Do Causal Powers Drain Away?" (Block), 124
Dopamine, 36, 64

Economic freedom, free will versus, 20–21
Edwards, Jonathan, 22
Einstein, Albert, 2, 41, 87, 128, 177n10
Eliminative materialism, 35–37, 59
Empiricism, constructive, 176n33
Endorsement, intentional, 110–111, 180–181n34
England, Deborah C., 29
Enlightenment, determinism in, 2–3, 38–40, 87
Epiphenomena, 44, 114

Epiphenomenalism: causal exclusion argument, 44–45; cause and effect, 114–118; epiphenomena, 44, 114; mental causation and, 43–48, 114, 118–123; neuroscientific support for, 46–48; premises of, 43–44, 113–114, 149
Equivalence of properties, 65–66
Expected utility, maximization of, 62
Explanatory necessity, 57–58
Extensional contexts, 175n21

"Fast" reasoning, 36
Feature binding, 64
Field, Hartry, 129
First-person consciousness, 11, 90
Fitness to be held responsible, 18, 154–156
Fodor, Jerry, 71
Folk science: causation as, 183n22; concept of, 37
Frankfurt, Harry, 168–169n29
Freedom, free will versus other notions, 20–21, 167–168n25, 168n26
Freedom House, 20
Freedom of Will (Edwards), 22
Free Will (Harris), 1
Free will, definition of, 15–17, 22–27
Free-will capacity, 27–29
Free-will emergentism, 10, 151. *See also* Compatibilist libertarianism
Free-will intuitions, 15–17, 165n7
"Free won't," 145
Fundamental level of reality, 124–126

Game theory, 61–62, 101
"Garden of Forking Paths," 79
Gazzaniga, Michael, 46, 114
Genuine causal overdetermination, 119, 182n8, 182n11
German Civil Code, 19
Gilboa, Itzhak, 179n24
Goal-seeking behaviour. *See* Intentional agency
Green, Thomas Andrew, 19
Grounding, 162n8
Groundwork for the Metaphysics of Morals (Kant), 165n8
Group agency, 154–156
Gyrus, risk-taking behaviour and, 36

Habermas, Jürgen, 167n22
Haggard, Patrick, 145
Halpern, Joseph, 131
Harari, Yuval Noah, 3–4
Hard incompatibilism, 151
Harris, Sam, 1, 3, 19, 46, 114
Haselager, Pim, 170–171n8
Haynes, John-Dylan, 47–48, 144
Heat equation, 55
Hidden-variables interpretations, 42
Higgs boson, 74, 102
Higher-level phenomenon: agential indeterminism as, 89–97; causation, 124–132; criteria for, 161–162n7; free will as, 4–6, 150; intentional agency as, 64–74; multiple realizability of intentional properties, 69–74, 91; nomologically possible histories, 94–97; obstacles to reduction, 64–69, 175n21; supervenience, 73–74, 91, 161–162n7, 162n8, 176n30
Histories: observational equivalence, possibility of, 178n17; physical (lower-level), 93–95; psychological (higher-level), 94–97
Hoefer, Carl, 12, 43, 163–164n18
Hume, David, 12, 115–117, 172n1
Humean compatibilism, 12, 115–117, 172n1
Hurley, Susan, 85–86

Imagined possibilities, 102–103
Impulsive behaviour, 36
Incompatibilism: approaches to, 163–164n18; Basic Argument for, 171n17; conflation of levels and, 12; hard, 151; illusion of, 12; libertarianism, 151; premises of, 151. *See also* Compatibilism
Indeterminism: challenge from, 107–111; human agency and, 97–103, 178–179n18; intentional endorsement, 110–111, 180–181n34; as prerequisite to free will, 152; quantum-mechanical, 43, 171–172n22; randomness, 43, 107–111; surface-level, 41–42. *See also* Agential determinism; Physical determinism
Indeterministic histories: physical (lower-level), 93–95; psychological (higher-level), 94–97
Inevitability, determinism and, 177–178n13
Inference to the best explanation, 58, 102, 117–118, 120, 173–174n14
Inferential statistics, causation in, 131
Inferior parietal lobule, 36
Input-output system, intentional agents as, 50–51
Insanity defense, 18, 23
Instrumentalism, 75–77, 101–102
Instrumental rationality, 51–52, 67
Intentional agency, 49–50; agency incompatibilism, 179n23; agenticity, 170n3; agentive modalities, 179n20; in animals, 58–62, 173–174n14, 173nn11–13; and capacity for free will, 27–29; case for realism about, 8–9, 74–77; causal control and, 25–26, 169n33; conditions for, 50–51; definition of, 22–23, 50–52, 53; higher-level nature of, 4–5, 64–74, 150, 175n21; indispensability of agency ascriptions, 58–64; instrumental rationality in, 51–52, 67; intentional actions, 52–54; intentional endorsement, 110–111, 180–181n34; intentional stance, 55–56, 154, 155; intentional states, 52–54, 172n4; interpretivism, 56–58, 173n10; maximization of expected utility, 62; multiple realizability of intentional properties, 69–74, 91; naturalistic ontological attitude applied to, 8–9, 74–77; neuroscientific challenges to, 36, 47–48, 170–171n8; in nonbiological entities, 154–156; overascriptions of, 34–38; predictive power of, 60–62; randomness and, 107–111; structure-agency debate, 62–63, 174n16; test for, 55–58. *See also* Materialism
Intentional endorsement, 110–111, 180–181n34
Intentionality, 33–35, 53, 62, 67. *See also* "Aboutness"; Intentional agency
Intentional stance, 55–56, 154–155
Intentional states, 51, 52–54, 172n4
Interpretivism, 56–58, 173n10
Interventionism, 128, 184n26
Intoxication, free will and, 28–29
Intuitions about free will, 15–17, 165n7
Irresistible impulse, 18
Ismael, Jenann, 12, 131, 163–164n18

Jackson, Frank, 133

Kahneman, Daniel, 36, 100, 179n22
Kane, Robert, 12; indeterminism, 43, 109–110, 163–164n18; self-formation, 180n32, 181n34

Kant, Immanuel, 17, 43, 165n8
Kenny, Anthony, 12
Kim, Jaegwon: causal exclusion argument, 44–45, 118–123, 182n7, 184n33; "fundamental level of reality," 124–125
Knobe, Joshua, 169n33
Kratzer, Angelika, 103–106

Laplace, Pierre-Simon, 2, 40, 87
Laplace's demon, 40, 87
Law, presupposition of free will in, 17–20, 23, 165n8, 166n10, 167n22
Levels: concept of, 5–7, 89–90, 161n4; conflation of, 12; lower-level phenomenon, 5–6, 93–95, 161–162n7; psychological, 89–90; supervenience, 73–74, 91, 161–162n7, 162n8, 176n30. *See also* Higher-level phenomenon; Histories
Lewis, David, 131
Liability, 166n10
Libertarian compatibilism, 12, 164n19
Libertarianism: definition of, 17; libertarian compatibilism, 12; political meaning of, 9–10; premises of, 26, 151; supervenient, 162n11. *See also* Compatibilist libertarianism
Libet, Benjamin, 47, 121–122, 141–147
Libet experiments, 47, 121–122, 141–147
Linking thesis, 88–89
Locke, John, 9
Lower-level phenomenon: criteria for, 161–162n7; definition of, 5–6; nomologically possible histories, 93–95
Luther, Martin, 24, 109, 111

Macroeconomics, causation in, 134–135
Maier, John, 179n20
Materialism: eliminative, 35–37; human agency in, 32–38; predictive power, failure of, 60–62; premises of, 31–32, 149. *See also* Intentional agency; Radical materialism; Reductive materialism
Maximization of expected utility, 62
Meaningful content, attitude towards, 53–54, 67
Mechanics, Newtonian, 39–41, 55, 127
Melden, A. I., 12
Mele, Alfred, 12, 109, 143, 163–164n18
Mental causation: causal exclusion argument, 44, 118–123; as difference making, 136–137; epiphenomenalist thesis, 43–48, 114; higher-level nature of, 150; possibility of acting otherwise, 168–169n29
Menzies, Peter, 11, 131, 132
Microtubules, 171–172n22
Modal interpretation of alternative possibilities, 81–86
Modalities, agentive, 179n20
Moore, G. E., 83–84
Moral responsibility: and capacity for free will, 27–29; morally responsible agents, 154; neuroscientific explanations of, 36; paleo-compatibilism and, 163n17; possibility of acting otherwise, 168–169n29; presupposition of free will, 17–20, 165n8, 166n10, 166n21, 167n22. *See also* Alternative possibilities
Motivational states, 51, 52–54
Multiple realizability of intentional properties, 69–74, 91, 95, 97, 137, 175n27

Nahmias, Eddy, 12, 144–147, 163–164n18
Naturalistic ontological attitude, 8, 74–77, 102
Naturalistic realism, 74–77
Neural plasticity, 71
Neuronal readiness potentials, observation of, 47, 141–147
Neuroscience: as complement to intentional explanations, 64, 170n4; conscious intention, 47–48, 64; feature binding, 64; intentionality and, 36, 170–171n8; neural encoding of beliefs, 71; neural plasticity, 71; neuronal readiness potentials, observation of, 47, 141–144; quantum indeterminacies and, 171–172n22; risk-taking behaviour, 64. *See also* Lower-level phenomenon
Newton, Isaac, 2, 39–40, 55, 127
Nomologically possible histories. *See* Histories
Nonbiological entities, free will in, 154–157
Nondomination, 167–168n25
Nonhuman organisms: free will in, 153–154, 185n8; intentional agency of, 58–62, 173–174n14, 173nn11–13
Nonidentity principle, 120–121
Nonretributivist punishment, 20
North Korea, social freedom in, 20
Norton, John D., 183n22
"On the Notion of Cause" (Russell), 126–127
Nozick, Robert, 9

Objects of beliefs, 172n4
Observational equivalence, 178n17
Old Testament, free will in, 2
Organized collectives, agency in, 154–156
Overdetermination, causal, 46–47, 119

Paleo-compatibilism, 12, 163n17
Pearl, Judea, 131
Penrose, Roger, 171–172n22
Pereboom, Derk, 19
Pettit, Philip, 133, 154
Phenomenology, 11
Photon behaviour, indeterminism in, 41–42
Physical determinism: agency incompatibilism, 179n23; compatibility with agential indeterminism, 89–97, 152; definition of, 2–3, 87; in Enlightenment worldview, 2, 38–40, 87; hidden-variables interpretations, 42; inevitability and, 177–178n13; Laplace's demon, 40; linking thesis, 88–89; physics, laws of, 39–41; premises of, 2–4, 26, 38–39, 79, 86–87, 149; probability and, 178n17; randomness, 107–111; randomness and, 43; thought experiment, 157–158. *See also* Alternative possibilities; Lower-level phenomenon
Physicalism. *See* Materialism
Physical stance, 56
Physical states, 53
Physics, laws of: causation and, 126–129; determinism and, 2–4; fundamental level of reality in, 124–126; hidden-variables interpretations, 42; indeterminism in, 41–43, 102, 171–172n22; lower-level nature of, 5–6; Newtonian, 38–39, 127; quantum mechanics, 41–43, 171–172n22, 176n1; relativity, theories of, 41,

Physics, laws of (*continued*) 128; string theory, 125; unobservables, 75–76, 101–102
Point particles, 127
Political freedom, free will versus, 20–21
Possibilities, alternative. *See* Alternative possibilities
Preference maximization theory, 99–100, 179n22
Presupposition of free will, 17–20, 165n8, 166n10, 166n21, 167n22
Price, Huw, 131
Probability, determinism and, 178n17
Projectivist views of causation, 116
Properties: multiple realizability of, 69–74, 91; obstacles to reduction, 67–68; token reduction, 72–73; type reduction, 72–73
Prospection, 165n7
Psychological level, 89–90
Psychology: eliminative materialism thesis, 35–37; indeterminism in, 97–103, 178–179n18; neuroscience as complement to, 170n4; psychological level, 89–90
Punishment, 19–20
Putnam, Hilary, 71, 76
Pyysiäinen, Ilkka, 35

Quantum mechanics: hidden-variables interpretations, 42; indeterminism in, 41–43, 171–172n22, 176n1

Raatikainen, Panu, 131
Rabinowicz, Wlodek, 11, 110–111
Radical materialism: eliminative materialism thesis, 35–37; human agency in, 32–38; predictive power, failure of, 60–62; premises of, 31–35, 149; reductive materialism versus, 37–38. *See also* Intentional agency
Radioactive decay, indeterminism in, 41–42
Randomness, 43, 107–111
Rational choice theory, 101, 110, 179n24
Rationality, instrumental, 51–52
Realism, case for, 8–9; alternative possibilities, 97–103; causal control, 132–140; fundamental level of reality, 124–126; intentional agency, 8–9, 74–77
Realist views of causation, 116
Realizability of intentional properties, 69–74, 91
Realization insensitivity, 137
Reasoning, causal, 114–118
Reducibility of properties, 65–66; multiple realizability of intentional properties, 69–74; obstacles to, 64–69, 175n21; token reduction, 72–73; type reduction, 72–73
Reductionist views of causation, 116
Reductive materialism: animals, agency of, 59; obstacles to reduction, 64–69, 175n21; premises of, 37–38; realizability of intentional properties and, 69–74. *See also* Intentional agency
Relativity, theories of, 41, 128
Representational states, 51, 52–54
Responsibility: and capacity for free will, 27–29; concept of, 2; fitness to be held responsible, 18, 154–156; intentional endorsement and, 110–111, 180–181n34; presupposition of free will, 17–20, 154, 165n8, 166n10, 167n22. *See also* Intentional agency

Retributive punishment, 19
Risk taking, neuroscientific explanations of, 36, 64
Robots, free will in, 156–157
Roskies, Adina, 12, 131, 138, 163–164n18
Rovelli, Carlo, 164–165n20
Russell, Bertrand, 126–127

Sartre, Jean-Paul, 19
Searle, John, 67
Sebass, Gottfried, 165n2
Self-conception, 2, 17–20
Self-formation, 110, 180n32, 181n34
Shermer, Michael, 170n3
Siderits, Mark, 12
Sleepwalking, free will and, 28
"Slow" reasoning, 36
Smilansky, Saul, 166n21
Social freedom, free will versus, 20–21, 167–168n25, 168n26
Social obligations, as constraints on action, 17
Special sciences, cause and effect reasoning in, 129–132
Stances: design, 56; intentional, 55–56; physical, 56
State coercion, social freedom and, 20–21, 167–168n25
States: intentional, 51, 52–54, 172n4; physical, 53
Statistics, causation in, 131
Steward, Helen, 12; agency incompatibilism, 163–164n18, 179n23; on alternative possibilities, 101; intentional agency of animals, 174n13, 185n8; interpretivism, 173n10; on neuroscientific challenges to free will, 36, 171n9
Stochastic systems, 93, 178n16
Stoicism, 19
Strawson, Galen, 109
Strict liability, 166n10
String theory, 125
Structure-agency debate, 62–63, 174n16
"Sufficient cause," 139–140, 184n33
Supernaturalism, 46–47, 119
Supervenience: concept of, 7, 73–74, 120–121, 161–162n7, 162n8, 176n30; physical and agential phenomena, 91, 94, 138, 142
Supervenient libertarianism, 162n11. *See also* Compatibilist libertarianism
Surface-level indeterminism, 41–42

Temporal succession, 128
Test of agency, 55–58
Theories of relativity, 41, 128
Third-person perspective, 11
Token reduction, 72–73
Tversky, Amos, 100, 179n22
Type reduction, 72–73

Uithol, Sebo, 170–171n8
Unobservables, 75–76, 101–102

Variables, control, 138, 146
Vihvelin, Kadri, 12, 171n17

Walter, Henrik, 165n2, 169–170n35
Wegner, Daniel, 47
Werndl, Charlotte, 178n16
What Scientific Idea Is Ready for Retirement? (Coyne), 1
Whittle, Ann, 85
Woodward, James, 128, 131